W9-AWH-534

DISCOVERY

Champion of the Space Shuttle Fleet

Valerie Neal

Curator, Smithsonian National Air and Space Museum

Smithsonian
National Air and Space Museum

Washington, D.C.

In association with

ZENITH PRESS

Dedication

With appreciation to all the men and women who participated in the Space Shuttle program—especially those who designed, built, serviced, supported, and flew on *Discovery* and those who delivered this national treasure to the Smithsonian.

First published in 2014 by Zenith Press, an imprint of Quarto Publishing Group USA Inc., 400 First Avenue North, Suite 400, Minneapolis, MN 55401 USA

© 2014 Quarto Publishing Group USA Inc.
Text © 2014 Smithsonian National Air and Space Museum

All rights reserved. With the exception of quoting brief passages for the purposes of review, no part of this publication may be reproduced without prior written permission from the Publisher.

The information in this book is true and complete to the best of our knowledge. All recommendations are made without any guarantee on the part of the author or Publisher, who also disclaims any liability incurred in connection with the use of this data or specific details.

We recognize, further, that some words, model names, and designations mentioned herein are the property of the trademark holder. We use them for identification purposes only. This is not an official publication.

Zenith Press titles are also available at discounts in bulk quantity for industrial or sales-promotional use. For details write to Special Sales Manager at Quarto Publishing Group USA Inc., 400 First Avenue North, Suite 400, Minneapolis, MN 55401 USA.

To find out more about our books, visit us online at www.zenithpress.com.

Library of Congress Cataloging-in-Publication Data

Neal, Valerie.
  Discovery : champion of the Space Shuttle fleet / Valerie Neal.
    pages cm
  Includes index.
   ISBN 978-0-7603-4383-8 (hardcover)
  1. Discovery (Spacecraft)--History. 2. Space flights--United States--Chronology. I. Title.
  TL795.5.N45 2014
  629.44'1--dc23

                    2014004401

On the front cover: *Discovery* launches on its first mission with a six-person crew to deploy three communications satellites and conduct science experiments, August 30, 1984. *NASA*

On the back cover: Astronaut Carl E. Walz (foreground) holds a power ratchet tool and James H. Newman tests the portable foot restraint in preparation for the first Hubble Space Telescope servicing mission. *NASA*

All images on pages 29 through 119 courtesy of NASA

Acquisitions Editor: Elizabeth Demers
Project Manager: Madeleine Vasaly
Design Managers: James Kegley and Rebecca Pagel
Designer: Kim Winscher
Layout: Rebecca Pagel

Printed in China
10 9 8 7 6 5 4 3 2 1

CONTENTS

Missions

ACKNOWLEDGMENTS

Without holding him responsible for the contents of this book, I freely admit that a "biography" of *Discovery* would not have come to fruition so well without the unfailingly generous attention of Dennis R. Jenkins, my informal collaborator and "quality control officer." He is an engineer and manager whose long career has been focused on the Space Shuttle, and he knows as much as, and probably more than, anyone about these remarkable vehicles. An author in his own right with many aviation and aerospace books to his credit, Dennis is now completing a three-volume technical history of the Space Shuttle, a work even more definitive than his classic earlier volumes on the history of the Space Transportation System.

Over the years, Dennis and I became acquainted through our work on *Enterprise* and *Discovery*, he as engineer-manager and I as curator. In the midst of concentrating on his *magnum opus*, Dennis graciously read my manuscript, gently guided me toward clarity on any matters needing correction, and helped me sort out credible data when sources differed. One might think that the number of orbits or maximum altitude of a shuttle mission would be a cold, hard absolute, but no; reports sometimes vary, so the numbers become slippery, and he helped me reconcile such discrepancies. He also gave me access to selected documents and images in his reference collection. Fortunately, we are both sticklers for details and patient in the effort to get things right, and this book is much more factually reliable due to his collegial involvement. Thank you, Dennis.

Thanks to NASA for putting so much information online, making it possible to write a book like this at all hours of the day and night, on weekends and holidays, without the constraints of office hours. Press kits, news releases, mission status reports, mission summaries, post-flight videos, crew interviews, photographs, and much more are so readily accessible that one need hardly step away from the computer to do research. At times, I had up to ten NASA websites open as I was writing, fact-checking, and tracking down pesky details.

Mary J. (Jody) Russell, who works in the media resource center at NASA Johnson Space Center, deserves credit for her knowledge of the image collection and her impeccable customer service. She is a one-woman rapid response team in filling requests for high-resolution images that are not available online. Like Space Shuttle workers I have known, she always gives her best effort without delay. Margaret A. (Maggie) Persinger in the media archives at NASA Kennedy Space Center also assisted quite ably in locating certain desirable photos

The staff of the National Air and Space Museum Archives, both the archivists and the photographers, are always a helpful resource, and I tapped them for aid in illustrating the last chapter, *Discovery*'s Final Mission. From the millions of images in the collection, they can find the perfect one or take a new one. For this project, archivists Amanda Buel, Allan Janus, Melissa Keiser, and Jessamyn Lloyd, as well as photographer Dane Penland, merit special notice. Almost everyone else on the museum's staff became involved in planning and executing *Discovery*'s transfer and display; without their teamwork and enthusiasm, none of that would have happened.

Patricia J. Graboske, publications officer at the museum, offered this project to me and nurtured our relationship with the publisher. She, Lawana Bryant in the Smithsonian Office of Contracts, and Jo Ann Morgan in the Space History Department deftly handled all the administrative matters. Space history intern Lynn Atkin helpfully compiled mission and crew data for the mission summaries. The Zenith Press editorial and design teams assigned to this project turned their many talents to making Discovery's story into a beautiful book. As editorial director for Zenith Press, Erik Gilg was excited about

the book from the outset and promoted it enthusiastically. It has been a pleasure to work with all of them to publish *Discovery*'s story.

Finally, I offer heartfelt appreciation to some people who are farther removed from this project but nevertheless merit acknowledgment. First are the two men who hired me in 1980 and put me to work on shuttle and Spacelab missions for a decade before I came to the National Air and Space Museum. Edwin C. Pruett of Essex Corporation and Charles R. (Rick) Chappell, a scientist at NASA Marshall Space Flight Center, saw potential in a newly minted PhD historian and encouraged me to write on spaceflight, space science, and space technology. It became my life's work, an adventure I never anticipated. Then, working on NASA and museum projects brought me into contact with many scientists, engineers, managers, and astronauts over the years who were generous in sharing their knowledge. They are too numerous to name, but I trust they know that our conversations made me smarter.

If I do justice to *Discovery* here, it is because these knowledgeable friends and colleagues influenced my growth as a writer and curator in the Space Shuttle era.

DISCOVERY
AND THE
SPACE SHUTTLE
ERA

1

***Discovery* became the champion** of the Space Shuttle fleet not simply because it flew more missions—thirty-nine in all—than *Columbia, Challenger, Atlantis,* and *Endeavour.* It also served longer—twenty-seven years—spending altogether 365 days in space. Its flight history began in 1984 as the fleet was starting its busiest two years and ended in 2011 as the shuttle program wound down. Most distinctively, *Discovery* flew every type of mission and served every purpose for which the Space Shuttle was designed. *Discovery* had no rival in the variety of its missions and the range of "firsts" it attained.

Discovery's story is the full Space Shuttle story in microcosm. Its thirty-nine-episode narrative traces high and low points in the four-decade quest by the United States to make human spaceflight in Earth orbit routine, practical, economical, and safe. *Discovery* alone lifts the story from tragedy back to triumph as the return-to-flight vehicle after both shuttle accidents. Its flight history makes *Discovery* a robust icon for the entire shuttle era.

The Space Shuttle came into being in the 1970s to continue American spaceflight after the space race and landings on the moon. With no national appetite for an expensive grand venture—a space station or a mission to Mars—and with social problems at home, the United States settled on a new Space Transportation System (STS): a fleet of shuttles designed for missions in Earth orbit. Flying often on various types of missions, shuttles presumably would reduce the cost of human spaceflight and expand its purpose. If they proved successful, shuttles

On August 30, 1984, *Discovery* launched on its first mission with a six-person crew to deploy three communications satellites and conduct science experiments. *NASA* might later pave the way to a space station or deep-space expeditions.

The key element, often called the workhorse or space truck or spaceplane, was a reusable orbiter, large enough to carry both people and payloads and versatile enough to keep dreams alive for a more exotic future. Attached to twin solid rocket boosters and pumping liquid hydrogen and oxygen propellants from an enormous external tank into its three internal launch engines, the vehicle streaked from Earth to orbit in eight and a half minutes. Shedding the boosters and tank during ascent, the spacecraft operated in the altitude range of 115 to 400 miles (185 to 645 kilometers) on stays ranging from two to eighteen days. Covered with a novel thermal protection system made of tiles and blankets, the orbiter descended from space

The Space Shuttle "stack" included the orbiter, its external propellant tank, and two reusable solid rocket boosters. Here, *Discovery* ascends on the STS-114 return-to-flight mission in 2005. *NASA*

Discovery set records in number of missions flown, total time and distance in orbit, and total number of crew members. *NASA*

Discovery made its final touchdown on March 9, 2011, to end the STS-133 mission to the International Space Station. The reusable Space Shuttle orbiter operated as a launch vehicle, crew ship, cargo carrier, and glider. *NASA*

to land on a runway. After several weeks of servicing, the orbiter was ready for its next mission.

The reusable shuttle would enable humans to begin living and working in space on a routine basis and using space near Earth for practical purposes. In the heady early days of development, planners envisioned spaceflight service as regular as an airline, with a fleet of five or more orbiters launching from sites in Florida and California as often as once a week. This forecast proved overly optimistic for a variety of reasons.

The Space Transportation System was meant to serve all of the nation's launch needs for commercial, scientific, and national security access to space. The plan called for the shuttle to become the sole launch vehicle for all types of payloads. With more onboard engineers and scientists than pilots, shuttle crews offered retrieval of errant satellites, in-orbit servicing of balky or failed equipment, hands-on laboratory research, and the skills for assembly of large space structures. Although for various reasons a large customer base did not materialize, in its first decade the shuttle served the needs of government, industry, and the scientific community as planned.

Skeptics doubted the economic benefits of the Space Shuttle before it began service and continued to challenge the wisdom of this approach to spaceflight

throughout its history. Yet from the successful first launch in 1981 to the 1986 launch tragedy, the shuttle ramped up in frequency and duration of flights. Nine missions launched in 1985, and 1986 was to have been even busier, with three orbiters and fifteen launches at an average rate of more than one per month. Spaceflight was beginning to seem routine. After the January 1986 *Challenger* accident brought shuttle flights to a halt for almost three years, the schedule gradually built up to seven and eight missions per year in the 1990s. Launches continued with few pauses for seventeen years until the *Columbia* accident temporarily grounded the shuttle again.

Discovery made its debut as the shuttle program was gaining momentum. Its first mission, STS-41D in August–September 1984, was twelfth in the program's schedule. *Discovery* immediately entered service for satellite deliveries and national security missions. In fewer than two years on duty until the first accident, *Discovery* flew six times, including three consecutive missions, rapidly approaching *Challenger*'s record of nine flights in three years. The future champion was proving its mettle.

The heart of this book is a mission log that presents *Discovery*'s missions in chronological order for easy reference, but first it is helpful to look at its history thematically to see trends and evolution in the

During the 1980s, most *Discovery* missions deployed communications satellites. In this view from the STS 51I mission in 1985, an Australian satellite with attached boost motor rises from the payload bay. Boost stages sent the satellites to geosynchronous orbit. *NASA*

shuttle program. Although every shuttle mission had several objectives, missions generally were designated by their primary purpose or payload into distinct types: commercial, national security, servicing, science, Mir visits, and International Space Station assembly or logistics. *Discovery* flew multiple missions of each type.

Discovery's first role was to deliver commercial satellites to low Earth orbit, from which they were propelled by attached stages to more distant geosynchronous orbits. Five of *Discovery*'s first six missions served customers from the communications satellite industry, and its first two post-*Challenger* missions deployed NASA Tracking and Data Relay System (TDRS) satellites. Some of the commercial missions also

delivered a satellite for the U.S. Navy. For satellite deliveries, the shuttle truly served as a cargo truck; two or three satellites were packed in the payload bay to be released one at a time when the orbiter reached the proper altitude and alignment. *Discovery*'s first mission was the first shuttle flight to carry three satellites. In all, *Discovery* delivered sixteen communications satellites for the United States, Canada, Mexico, the Arab League, and Australia. They represented the growing market of non-spacefaring nations eager to join a global telecommunications network.

The commercial sector was crucial to the effort to make human spaceflight more economical and routine, and NASA's business plan depended on a growing

Discovery crews completed two of the five servicing missions that extended the life of the Hubble Space Telescope well beyond its planned ten years. In this view from the 1999 servicing mission, astronauts Steven L. Smith and John M. Grunsfeld, working at the end of the long robotic arm, have opened a bay to replace gyroscopes in the pointing and attitude control system. *NASA*

and active customer base for satellite deliveries and also for research projects. NASA cultivated commercial customers with attractive pricing and incentives, including the opportunity for a corporate payload specialist to join the crew. The first commercial payload specialist, Charles D. Walker of McDonnell Douglas, flew on *Discovery* twice and *Atlantis* once to conduct experiments in a potentially lucrative manufacturing process. Saudi Arabia and France placed payload specialists on *Discovery* to witness the deployment of their satellites, as did Mexico on *Atlantis*. Three members of Congress took advantage of this courtesy and persuaded NASA to put them on crews—Senator Jake Garn on *Discovery* in 1985, Representative William "Bill"

Nelson on *Columbia* in early 1986, and Senator John Glenn on *Discovery* in 1998.

Commercial payloads occasionally prompted another mission type: servicing. Twice *Discovery* crews combined deployments with retrievals for repairs or returns when satellites failed to reach their required orbits. The crew of *Discovery*'s second mission celebrated the first in-orbit satellite retrieval with a "2 Up, 2 Down" sign when they successfully deployed two satellites and picked up two others to bring home for refurbishing. Another *Discovery* crew brought an idle satellite into the payload bay, installed a new ascent motor, redeployed it, and watched it ignite on its intended path. These servicing episodes demonstrated

Discovery's science missions typically included retrievable research satellites and instruments mounted in the payload bay. This ATLAS 2 suite of instruments for atmospheric and solar physics investigations flew on the STS-56 mission in 1993. *NASA*

important crew skills and built crew experience for future projects, notably servicing the Hubble Space Telescope and assembling a space station.

Policy changes after the *Challenger* accident took commercial satellites off the shuttle and seriously eroded the commercial market for shuttle flights. Commercial experiments continued to fly as secondary payloads, but *Discovery's* role soon shifted from satellite delivery to other types of missions.

The Department of Defense reserved *Discovery's* third flight for the first dedicated, classified, national security shuttle mission, about which little is known beyond the names of the first all-military crew and the first U.S. Air Force payload specialist who was not in the NASA astronaut corps. The primary payload was presumed to be an electronic intelligence satellite.

Although the shuttle was designed and developed with national security needs in mind, the air force grew reluctant to rely on the shuttle and NASA as the sole launch provider, preferring instead to maintain its own capacity for assured access to space. Even before the *Challenger* accident and subsequent grounding of the fleet, the Department of Defense began to ease away from the shuttle. It completed its backlog of planned national security missions when flights resumed and then abandoned the shuttle except for occasional small, unclassified payloads.

From 1984 through 1992, *Discovery* flew four of the ten Department of Defense missions. The first two of these were strictly classified, but allegedly spy satellites were deployed for the National Reconnaissance Office. The other two missions were publicly linked to the Strategic Defense Initiative ("Star Wars"), one unclassified and the other partially cloaked in secrecy. Editorial cartoonists mocked the military missions by portraying the shuttle in disguise or as invisible, but the point was to question militarization of space and the place of secrecy in a public space program defined by its open conduct. The issue dissipated as the air force returned to its preferred rockets and the Space Shuttle moved on to almost exclusively civilian tasks.

In 1990 and again in 1997 and 1999, *Discovery* and its crews made history, first by deploying the heralded Hubble Space Telescope and then by returning twice to service, repair, and redeploy it. *Discovery* was not scheduled for the urgent first visit to install corrective optics, but it drew duty for the second and third of five servicing missions. Spacewalking teams updated the telescope with newer technologies and extended its life by repairing or replacing worn components. With skill and finesse, *Discovery* crews demonstrated the value of humans in space for efficient performance of complex tasks. In-orbit servicing benefited the astronomical community by adding years of continued telescope operations and also built the experience base for future large assembly projects such as the International Space Station.

On missions to the Hubble Space Telescope, the orbiter's name seemed especially apt. Namesake of the exploring ships of Henry Hudson and Captain James Cook, *Discovery* furthered the tradition of exploration and discovery through improved observation of the universe.

Discovery's primary occupation in the 1990s, however, was to support science. On ten missions during the decade, this orbiter carried satellites, observatories, or laboratories for scientific research. It also delivered the sun-circling explorer called Ulysses. NASA's science missions had several purposes: to exploit microgravity as a laboratory environment, to understand better the changes in humans and other organisms during long stays in space, to take advantage of the clear viewing

Discovery made the first and last of nine shuttle missions to Russia's Mir space station, seen here during the shuttle's final departure in 1998. *NASA*

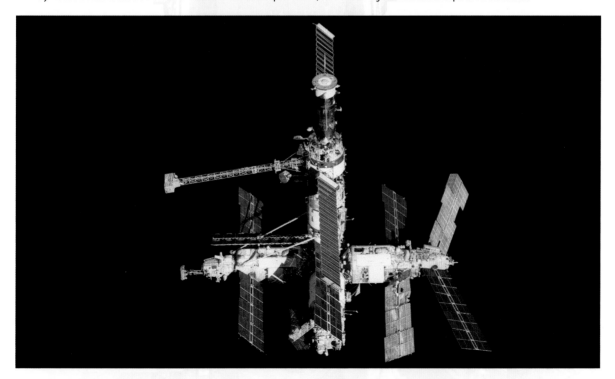

Discovery is seen approaching the International Space Station on its last mission, STS-133 in February 2011, making its thirteenth trip to the orbital outpost. *NASA*

of Earth and cosmos from above the atmosphere, and to pursue both basic and applied science in quest of benefits for people on Earth.

Discovery's science missions covered a range of disciplines—primarily Earth and atmospheric observations, materials processing, biology, and biomedicine. Several times it carried a Spacelab or SPACEHAB laboratory module in the payload bay, where scientists worked in shifts around the clock. Other times it carried a platform loaded with largely automatic devices, or it released and recovered a small, free-flying satellite for particular experiments. *Discovery* flew two "Mission

to Planet Earth" flights among its science missions in the 1990s.

Discovery opened 1995 with the first of nine shuttle missions to the Russian space station Mir in a cooperative program to prepare for International Space Station partnership. *Discovery* made the first Mir close approach and fly-around, and in 1998 it completed the final Mir docking mission, returning the last of seven American astronauts who alternately lived on the space station. For the Shuttle-Mir missions, the Russian and U.S. space agencies shared training, crews, orbital operations, and mission control

STS-114 commander Eileen M. Collins put *Discovery* through the rendezvous pitch maneuver, or backflip, for the first time upon approaching the International Space Station in 2005. Cameras and crew on the station surveyed the vehicle for signs of damage. *NASA*

responsibilities to set the stage for their roles as principal space station partners in the new century. This series of missions included a number of firsts, several of them on *Discovery*.

From 1999 on, all but one of *Discovery*'s last fourteen missions went to the International Space Station. *Discovery* made the first docking with the nascent station to prepare it for its first resident crew and paid its final visit to the station filled with supplies. Its primary roles in space station assembly were deliveries of long truss segments, the Harmony connecting node, Japan's Kibō laboratory module, and the Leonardo Multipurpose Logistics Module. Most of its missions also included space station crew exchanges. *Discovery* flew one more mission to the space station than the other orbiters; its only other mission in this period went to service the Hubble Space Telescope.

Twice in its history, *Discovery* stood poised on the launch pad to take American astronauts into orbit after a long pause in the Space Shuttle program. With the loss of *Challenger* during its tenth launch in 1986 and *Columbia* during its twenty-fifth reentry in 2003, *Discovery* became the return-to-flight vehicle. In fact, it bore that responsibility three times; the first two post-*Columbia* missions, both on *Discovery*, tested new procedures to detect and repair damage in orbit and thus reduce the risk of another fatal mishap. Executing their missions flawlessly, three *Discovery* crews and this spacecraft restored confidence in the Space Shuttle after profound losses.

During twenty-seven years in service, *Discovery* tallied a number of firsts in its operations and crews. It was the first orbiter to carry and deploy three satellites on a single mission; the first to rescue and return satellites from space; the only orbiter to fly four times in a year; the first and last to visit the Russian space station Mir; the first to dock with the International Space Station; the ship that flew the one hundredth

The crews of *Discovery*'s STS-120 mission (in red) and International Space Station Expedition 16 (in blue) pause from work to demonstrate weightlessness inside the newly installed Harmony node. This crew portrait, like most during the shuttle era, reflects the new demographics of spaceflight. It also captures the first time two women held simultaneous command of space missions: Peggy A. Whitson (lower row) and Pamela A. Melroy (middle row, center). *NASA*

shuttle mission; the only orbiter to fly three Hubble Space Telescope missions; the first to perform the rendezvous pitch maneuver (backflip in space) for heat shield inspection; and the first to carry the robotic arm's extended boom and sensors used by the crew to inspect the entire outer surface of the orbiter for damage.

Discovery's crews for its thirty-nine missions reflected the diversity of the astronaut corps in the Space Shuttle era. Its thirty-two commanders (six of whom commanded it more than once) included the first African American commander, another African American commander who later became NASA Administrator, the first female pilot, and both female commanders. The 184 astronauts, cosmonauts, and payload specialists who flew on *Discovery* included twenty-eight women and citizens of Canada, France,

Germany, Italy, Japan, Russia, Saudi Arabia, Spain, Sweden, Switzerland, and the United States. The first Asian American in space, the first Hispanic woman, the first Arab, the first two Canadian female astronauts, the first African American spacewalker, the first astronauts from Spain and Sweden, the first commercial and military payload specialists, the first two cosmonauts on shuttle missions, the only person to do a spacewalk underneath the orbiter, the first U.S. senator, and the only Mercury astronaut to fly on the shuttle all served on *Discovery*.

In 2004, President George W. Bush announced that the Space Shuttle program would end upon completion of the International Space Station, and NASA soon began planning to retire the orbiters. Each vehicle made a final flight in 2011 to put finishing touches on the station and bring stockpiles of supplies to meet crew needs for the foreseeable future without a large cargo vehicle for resupply flights. On its last trip in space, *Discovery* delivered supplies, a storage module, and the astronauts' assistant, Robonaut 2, to the station.

Discovery's final voyage into retirement signaled the end of the Space Shuttle era. It began in Florida and ended in the nation's capital on April 17, 2012. NASA assigned *Discovery* to the Smithsonian National Air and Space Museum, and when the orbiter was properly configured for public display, mounted it on the Boeing 747 Shuttle Carrier Aircraft for the journey to its permanent home. After a spectacular fly-over of the Washington, D.C., metropolitan area that brought crowds pouring out of homes, schools, and office buildings to witness the sight, the paired craft landed at Dulles International Airport in suburban Virginia.

Two days later, *Discovery* arrived at the museum's Steven F. Udvar-Hazy Center for a ceremonial transfer. Now the champion orbiter occupies the center of the museum's space hangar as if it landed in this place of honor. *Discovery* stands amid other historic rockets and spacecraft as a lasting icon of American achievements in space and the effort to make human spaceflight routine.

Discovery's arrival cruise over Washington, D.C., and surrounding areas on April 17, 2012, paid tribute to the nation's capital, where political debates and decisions shape the course of spaceflight. *NASA*

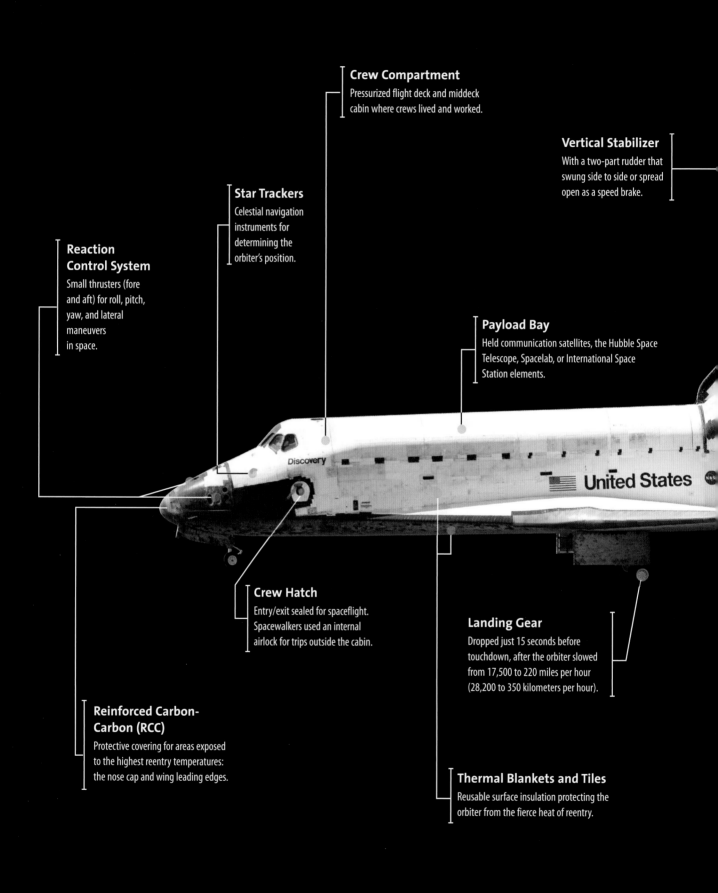

Crew Compartment
Pressurized flight deck and middeck cabin where crews lived and worked.

Vertical Stabilizer
With a two-part rudder that swung side to side or spread open as a speed brake.

Star Trackers
Celestial navigation instruments for determining the orbiter's position.

Reaction Control System
Small thrusters (fore and aft) for roll, pitch, yaw, and lateral maneuvers in space.

Payload Bay
Held communication satellites, the Hubble Space Telescope, Spacelab, or International Space Station elements.

Discovery

United States

Crew Hatch
Entry/exit sealed for spaceflight. Spacewalkers used an internal airlock for trips outside the cabin.

Landing Gear
Dropped just 15 seconds before touchdown, after the orbiter slowed from 17,500 to 220 miles per hour (28,200 to 350 kilometers per hour).

Reinforced Carbon-Carbon (RCC)
Protective covering for areas exposed to the highest reentry temperatures: the nose cap and wing leading edges.

Thermal Blankets and Tiles
Reusable surface insulation protecting the orbiter from the fierce heat of reentry.

DISCOVERY INSIDE AND OUT

Main Engines

Reusable rocket engines fired only for launch, draining the huge external propellant tank to reach orbit in just eight minutes.

OMS Pods

Two mid-size Orbital Maneuvering System engines for the shuttle's final push into orbit, changing altitude or velocity in space, and starting its descent home.

Elevons

Trailing edge flaps on the triangular delta wings, used for flight control in the atmosphere during descent to landing.

Although spaceplanes had been imagined for some time, before 1981 nothing like the Space Shuttle had ever flown in space. With wings and wheels, the shuttle orbiter resembled an aircraft but operated as a rocket during launch and a glider during the return from space. The wide delta-shaped wings enhanced maneuverability for the unpowered descent back to land, and the tall vertical stabilizer served as a rudder and speed brake. Three massive engines and associated plumbing filled the aft fuselage below the vertical stabilizer, and two bulbous pods at its base held smaller engines and thrusters for maneuvering in space, as did the nose. Comparable in size to a Boeing 737 or Douglas DC-9 passenger jet airliner, the Space Shuttle orbiter was the most technically complex spacecraft ever built.

Like a freight truck plying the highways, the orbiter had a crew cabin in front of a long cargo container—the payload bay that constituted the mid-fuselage. Satellites, laboratories and observatories, large experiments, and the structural beams and modules of the International Space Station rode here. Along the floors and walls of the payload bay, under protective blankets, lay miles of wiring and lines for the vehicle's electrical and fluid systems. The long, curved payload bay doors stayed open in orbit to expose the attached radiator panels.

A jointed mechanical arm, the Canadarm Remote Manipulator System (RMS), mounted along the port sill of the payload bay, served as a crane for transferring payloads and a mobile work platform for spacewalking astronauts. On missions after 2003, a straight arm extension equipped

STS-95 mission commander Curtis L. Brown stands at the aft flight deck workstation to maneuver the orbiter in space. *NASA*

with a cluster of cameras and sensors was mounted along the starboard sill for inspecting the spacecraft in orbit. Also mounted in the payload bay were communications antennas, cameras, floodlights, and the airlock-docking adapter needed for spacewalks and space station visits.

The crew cabin had two habitable levels: the upper flight deck and the middeck living quarters, with openings between them and a storage area below the middeck floor. The cabin normally accommodated up to seven people (and twice eight) with reasonable comfort and was considerably more spacious than previous capsules. The crew managed most of the orbiter's operations from the forward and aft flight decks. Pilot astronauts could maneuver the orbiter in space from either position. Crewmembers operated the arm remotely from an aft workstation, and astronauts on camera duty jockeyed for position at the ten upper-level windows.

While the flight deck was the hub of activity for ascent, descent, and in-orbit maneuvers, the middeck served as home for the usual activities of daily life—eating, sleeping, personal hygiene, exercise—and also as a workplace for certain experiments and housekeeping tasks. The middeck contained a small galley for preparing packaged foods, a waste management compartment (toilet), stowage lockers for crew

Shuttle astronauts considered the middeck their home in space. Its provisions were much like those for a camping trip. In this view of about half the middeck, STS-128 astronaut Christer Fuglesang from Sweden is near the galley and lockers, with the crew hatch window and toilet compartment behind him. *NASA*

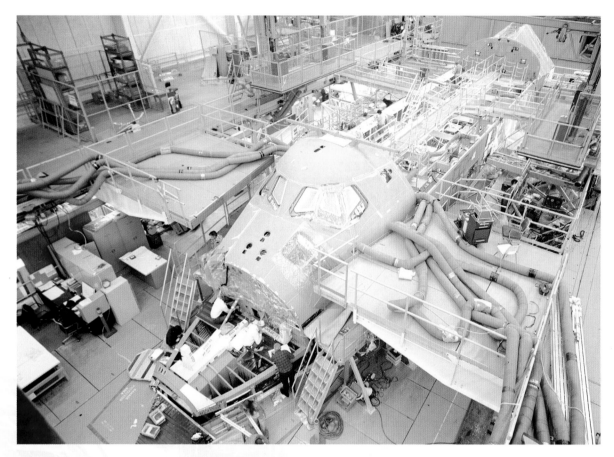

Assembly of *Discovery* began in Palmdale, California, in 1980 and was completed for a November 1983 delivery to Kennedy Space Center. *NASA*

equipment and scientific experiments, sleeping bags, and a hatch to the airlock used to exit the orbiter in space. The crew hatch on the port side of the vehicle held a small, round window.

The crew hatch afforded middeck entry and exit before and after flight but was otherwise sealed for flight unless the crew had to abandon ship in an emergency during ascent or reentry. This never happened. On the ground, the crew could escape, if necessary, by rappelling out of an overhead window or using an inflatable slide from the crew hatch. On the starboard side of the orbiter, rescuers could open a marked area into the middeck if the crew needed emergency assistance out of the vehicle.

Discovery was the fourth orbiter to be built (following *Enterprise*, *Columbia*, and *Challenger*) and the third to enter service in space. Designated as orbiter vehicle OV-103 and delivered to Kennedy Space Center, Florida, in 1983, *Discovery* was the first production orbiter, an improved vehicle refined by lessons learned in building and flying the developmental orbiters. *Discovery* weighed almost 7,000 pounds less than its predecessors, thanks to changes in the orbiter's structural design, and thus could carry more payload weight. It also was the first orbiter on which flexible blankets replaced most of the white tiles forming part of the reentry heat shield.

Thermal Protection System

Much of *Discovery*'s history is suggested in the discolored mosaic of tiles, blankets, and carbon panels that cover the vehicle. Together they formed a reusable, repairable, lightweight heat shield that permitted the vehicle to return from space time and again—thirty-nine times in all.

The orbiter's descent through the atmosphere at Mach 25 generated temperatures up to 3,000°F (1,650°C). To survive this reentry furnace, the thermal protection system kept the vehicle's aluminum airframe and skin below 350°F (175°C). Reinforced carbon-carbon panels shielded the nose cap and wing leading edges from the most extreme heat. Silica tiles and flexible silica blankets that contained more

Discovery's surface is streaked, faded, dinged, and discolored—the hallmark of traveling to space and back again and again. *Photo by Dane Penland, National Air and Space Museum (NASM 2013-02524)*

OV-103 *DISCOVERY* SPECIFICATIONS

Manufacturer: Rockwell International, prime contractor
Forward Fuselage and Crew Cabin: Rockwell International
Midfuselage: General Dynamics
Payload Bay Doors: Rockwell International
Wings: Grumman
Aft Fuselage: Rockwell International
Vertical Stabilizer: Fairchild Republic
Main Engines: Rockwell International, Rocketdyne Division
Length: 122 feet (37 m)
Wingspan: 78 feet (25 m)
Height (on wheels): 57 feet (17 m)
Payload Bay: 60 feet x 15 feet (18 x 5 m)
Orbital Altitude Range: 115 to 400 statute miles (185 to 643 km)
Maximum Cargo to Orbit: 65,000 pounds (29,400 kg)
Heaviest Weight at Launch: STS-124: 269,123 pounds (122,072 kg)
Weight (current): 161,325 pounds (73,175 kg)

DISCOVERY'S OPERATIONAL RECORD

Missions: 39
Cumulative Time in Orbit: 365 days
Approximate Total Distance Traveled: 148 million miles (238 million km)
Crew Size: 5, 6, or 7 per mission
Shortest Mission: 3 days, 1 hour, STS-51C (1985)
Longest Mission: 15 days, 2 hours, STS-120 (2007) and STS-131 (2010)
Communications Satellite Missions (1984–89, 1995): 8
Department of Defense Missions (1985–92): 4
Scientific Missions (1990–98): 9
Hubble Space Telescope Missions (1990, 1997, 1999): 3
Mir Missions (1995, 1998): 2
International Space Station Missions (1999–2011): 13
Day/Night Launches: 29 and 10
Day/Night Landings: 31 and 8
Landings in Florida/California: 24 and 15

air than solid material covered the rest of the orbiter; they radiated heat quickly and absorbed heat slowly during the 45-minute descent. Black tiles bore the brunt of heat up to 2,300°F (1,260°C) on the underside, nose, window frames, edges of the stabilizer and wing trailing edges, and areas around all the engine and thruster nozzles. White tiles and blankets covered the parts of the vehicle that "felt" lower temperatures (less than 1,200°F; 650°C)—the upper and side surfaces of the wings, fuselage, and stabilizer—which were partly protected by the orbiter's nose-high reentry angle.

Discovery sports about twenty-four thousand tiles and blankets, each with a unique part number, custom-fitted in size and shape to its specific location on the orbiter and individually installed by hand. On average, about one hundred tiles were replaced after each mission. About eighteen thousand of *Discovery*'s original black tiles remain in place, now streaked and aged to shades of gray in contrast to the glossy replacements. Chalk-white streaks, thin and straight as if airbrushed from the corners of the tiles, actually record the heat flow pattern over the orbiter during its descents. Small white circles mark the spots for waterproofing injections to keep the porous tiles from absorbing moisture. The now-blemished thermal protection armor is the badge of *Discovery*'s multiple homecomings.

Canadarm Remote Manipulator System

The most important aid for spacewalking astronauts, other than the protective Extravehicular Activity (EVA) spacesuit, was the Canadarm, also known as the Remote Manipulator System (RMS) or robotic arm. This 50-foot-long (15 meters) jointed arm extended the astronauts' strength and reach. It was used like a crane to move very large items in and out of the orbiter's payload bay. It also served as a mobile platform to position astronauts precisely where they needed to be while working outside.

The Canadian Space Agency and Canadian Commercial Corporation supplied several robotic arms for the shuttle and space station programs. One of

Discovery's thermal protection tiles were crucial for a reusable spacecraft; now they record the orbiter's history of returns from space. *Photo by Dane Penland, National Air and Space Museum (NASM 2013-02986crop)*

those, serial number 202, flew on *Discovery*'s last six missions, all to the International Space Station, and retired with *Discovery* after being used on fifteen shuttle missions and four orbiters from 1994 through 2011. Today, this arm is displayed beside the orbiter.

The robotic arm was actually a tele-robot, remotely operated by an astronaut at the aft flight deck control station inside the orbiter. The arm had shoulder, elbow, and wrist joints plus a snared end effector "hand" for grappling and a foot restraint that attached to the end. The operator moved the arm by using two hand controllers, aided by data displays, views from television cameras mounted on the arm, and line-of-sight monitoring through the orbiter's aft and overhead windows. The job demanded intense concentration and excellent hand-eye coordination.

Modernization

Discovery on display today is not exactly the same *Discovery* that first flew in 1984. During the era of shuttle operations, each of the orbiters underwent periodic refurbishment to update selected equipment, in addition to regular repairs and servicing between missions.

In 1992, after its fourteenth flight, STS-42, *Discovery* received nearly eighty modifications while in the servicing bay at Kennedy Space Center. These updates included installation of a drag chute below the vertical stabilizer to help slow the returning orbiter after touchdown.

In 1995–96 after STS-70, its twenty-first flight, *Discovery* spent nine months off duty to undergo a major upgrade back in the assembly plant in Palmdale, California. Almost one hundred changes were made, the most prominent being removal of the middeck airlock and installation of a new airlock in the payload bay. Many others were not as readily visible but were important to improve the vehicle's performance and durability.

In 2001, *Discovery* again entered major maintenance for installation of modern electronic flight deck instrumentation, the "glass cockpit." An array of eleven screens gave the crew multicolor displays of essential flight and systems data. That upgrade occurred after *Discovery*'s thirtieth flight, STS-105.

After the *Columbia* tragedy in 2003, *Discovery* was selected as the return-to-flight orbiter because it had been most recently upgraded. It remained under modification into 2004 and then flew two missions in a row, STS-114 in 2005 and STS-121 in 2006. *Discovery* was the first orbiter to receive post-accident modifications, including installation of the manipulator arm extension and sensor package for in-flight inspections of the vehicle. That boom remains installed inside *Discovery* today.

When the Space Shuttle program ended in 2011, *Discovery* and the other orbiters were not outdated. Through regular servicing between missions and periodic upgrades, these remarkable spacecraft had not yet reached their expected service life and might well have continued flying.

Right: The "glass cockpit" upgrade replaced many old electromechanical gauges and meters with an electronic system that presented information in color graphics on liquid crystal display screens. *NASA*

Opposite: Seen from the International Space Station on *Discovery*'s last flight, the jointed Canadarm with a straight boom extension rests above the payload bay. All mission crews after the 2003 loss of *Columbia* used this extension and its attached sensor system to inspect the entire vehicle exterior for damage. *NASA*

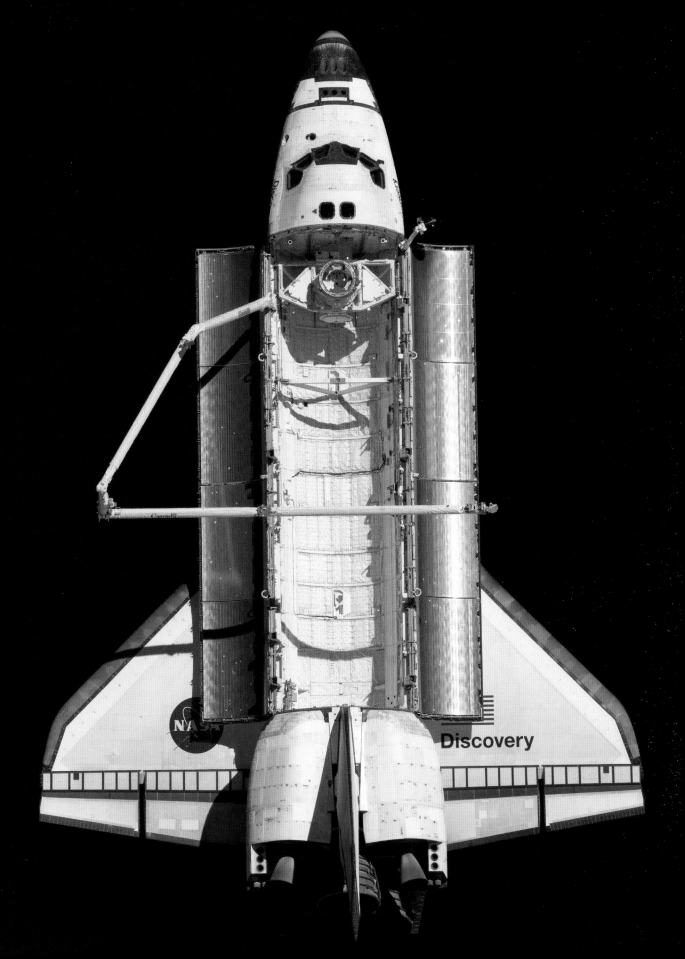

STS-51 astronaut Carl E. Walz (fore) holds a power ratchet tool, while James H. Newman tests the portable foot restraint in preparation for the first Hubble Space Telescope servicing mission.

MISSIONS

The following account of *Discovery*'s thirty-nine missions differs from a traditional logbook by being more narrative than numerical. Important information for each mission appears in a sidebar: launch and landing dates, flight duration, number of orbits, orbit inclination, maximum altitude, and number of extravehicular activities (EVAs), as well as landing site and crew roster. These mostly quantitative data form the skeletal record of the mission.

The body and personality of each shuttle mission reside in its story. Any single mission warrants an entire book, and someday the full story of each one may be told in detail. The aim here is to present summaries that highlight the essential character of each mission and also place it in context with other missions and developments in the Space Shuttle program. A narrative thread links *Discovery*'s missions to the broader history of the shuttle era, the other orbiters, and the International Space Station.

Come along for the rides, all thirty-nine of them.

6 : 0 : 56
DAYS HRS MINS

0 : 0
HRS MINS
0

201 M
175 NM
(324 KM)

97 | 28.5°

EDWARDS AIR FORCE BASE

COMMANDER

- Henry W. Hartsfield Jr., USAF, his 2nd of 3 flights

PILOT

- Michael L. Coats, USN, his 1st of 3 flights, all on *Discovery*

MISSION SPECIALISTS

- Judith A. Resnik, electrical engineer, her 1st of 2 flights

- Steven A. Hawley, astronomer-astrophysicist, his 1st of 5 flights, 3 on *Discovery*

- Richard M. (Mike) Mullane, USAF, aeronautical engineer, his 1st of 3 flights

PAYLOAD SPECIALIST

- Charles D. Walker, McDonnell Douglas, test engineer, his 1st of 3 flights, 2 on *Discovery*

Discovery's Debut Mission

Discovery entered service in 1984 as the third orbiter in the Space Shuttle fleet. *Columbia* and *Challenger* had already flown a total of eleven missions as America's "space trucks." *Discovery*'s first mission followed suit as the crew deployed, for the first time, *three* communications satellites, but it also signaled how the shuttle could serve as more than a delivery vehicle.

Discovery's first mission began with drama—three launch delays, the first on-pad engine shutdown (just 4 seconds before liftoff), a related fire on the launch pad, and rollback from the pad for a major payload shuffle—but after launch everything went smoothly. All but mission commander Henry Hartsfield were first-time flyers, and mission specialist Judy Resnik became the second American woman in space.

The crew released one satellite a day to start the mission: first an SBS for Satellite Business Systems, then a LEASAT (SYNCOM) for the U.S. Navy, and finally a TELSTAR for AT&T. After each deployment, the orbiter moved away before the time-delayed ignition of a boost motor sent the satellite to its ultimate orbit, some 22,300 miles (35,400 kilometers) high above the equator. The SBS and TELSTAR left the payload bay spinning like tops. The largest of the three, LEASAT was the first wide-body satellite designed for launch from the Space Shuttle.

Both the crew portrait and mission patch depicted the orbiter with an odd feature that looked like a tower rising from the payload bay. It was a 10-story, 13-foot wide, lightweight solar array, accordion-pleated for compact stowage—at that time the largest structure ever deployed in space. From the aft flight deck, the crew extended and retracted the array several times to observe its operation and stability. This new technology experiment used the shuttle as a test bed for evaluating large structures needed for a future space station.

The emblem for *Discovery*'s first voyage in space pays tribute to its namesake ships of exploration, with stars to indicate the twelfth shuttle mission.

Front (left to right): Richard M. (Mike) Mullane, Steven A. Hawley, Henry W. Hartsfield Jr., and Michael L. Coats. Behind: Charles D. Walker and Judith A. Resnik.

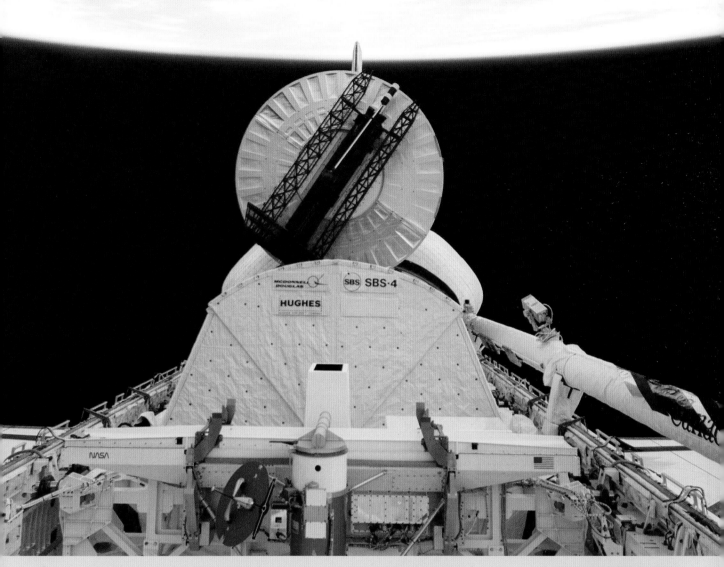

Before release, the 20-foot-long (6 meters) LEASAT (SYNCOM) communications satellite lay behind the SBS satellite in the payload bay with its antennae folded against the 14-foot (4.2-meter) diameter top. A spring-loaded trigger pitched it out like a slow-motion Frisbee.

NASA offered its corporate customers the opportunity to send their own payload specialist to conduct research in space. The first non-astronaut to fly on the shuttle, test engineer Charlie Walker of McDonnell Douglas, tended to a materials processing experiment of interest to the pharmaceutical industry. This program showcased the shuttle's usefulness for commercial research into the feasibility of manufacturing in space.

Discovery checked out well in space. However, one surprise drew attention: An icicle about 2 feet long and a foot in diameter, composed of wastewater and urine, jutted out from a dump port just beyond the crew hatch, where it threatened to damage the open payload bay door. The crew rotated the orbiter to expose that side to direct sunlight and reduce the ice mass, then tapped it gently with the robotic arm to break it loose.

The successful STS-41D mission confirmed the shuttle's versatility as a delivery vehicle, technology test bed, and research environment. More than twenty of *Discovery*'s thirty-nine missions also involved deliveries, but this orbiter began its career serving multiple purposes at once.

The lightweight, collapsible solar array, seen through an overhead window, extended 102 feet (32 meters) from the payload bay.

213 M
185 NM
(343 KM)

127 | 28.5°

KENNEDY
SPACE CENTER

2

COMMANDER

- Frederick H. (Rick) Hauck, USN, his 2nd of 3 flights, 2 on *Discovery*

PILOT

- David M. Walker, USN, his 1st of 4 flights, 2 on *Discovery*

MISSION SPECIALISTS

- Anna L. Fisher, physician, her only flight, RMS operator

- Dale A. Gardner, USN, engineer, his last of two flights, 2 EVAs

- Joseph P. Allen IV, physicist, his last of two flights, 2 EVAs

"Two Up, Two Down"

Discovery's next mission, barely two months after its first, delivered two communications satellites and rescued two others that had failed to reach their intended orbit. *Challenger* was originally scheduled for this first space salvage mission, but it had to stay in the servicing bay for extensive tile repairs after its April 1984 flight.

Each day of the eight spent in orbit had a single primary task. On the second and third days, the crew deployed satellites, the first called Anik for Telesat Canada, and then a second LEASAT (SYNCOM) for the U.S. Navy. These largely automated releases went as smoothly as those on the previous mission.

On the fourth day, the crew prepared to retrieve two stranded satellites that were deployed on *Challenger*'s STS-41B mission but were rendered ineffective when their boost motors misfired. The satellites had been remotely maneuvered to the shuttle's altitude and parked there for pickup. Commander Rick Hauck and pilot Dave Wilson guided *Discovery* to rendezvous with each of them. Then the crew began a task that had been done only once before, on *Challenger*'s STS-41C mission: capturing a satellite and bringing it into the payload bay.

On days five and seven, with a day of further preparation between, two mission specialists ventured out on spacewalks to capture first an Indonesian communications satellite, Palapa, and then Western Union's Westar. Joe Allen and Dale Gardner took turns wrangling the satellites back to the orbiter. Each time, one of them donned the Manned Maneuvering Unit (MMU) propulsion backpack and flew out with a "stinger," a long capture device that he inserted into the booster nozzle as a handle to halt the satellite's slow rotation and steer it back to the shuttle.

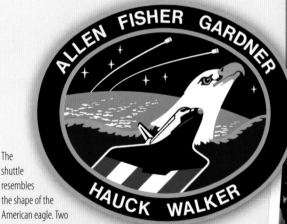

The shuttle resembles the shape of the American eagle. Two satellites enter orbit, and five stars represent the crew.

Front: Frederick H. (Rick) Hauck. Behind (left to right): Dale A. Gardner, David M. Walker, Anna L. Fisher, and Joseph P. Allen IV. An American eagle appeared as the STS-51A mascot in both the crew portrait and mission emblem.

Gingerly moving the satellites into the payload bay became a one-woman, two-man job. Anna Fisher operated the Remote Manipulator System (RMS) arm from the aft flight deck control station to grasp each massive satellite and bring it into the payload bay. She intended to place each one very carefully into its cradle for the trip home, but a hardware misfit forced the crew into "Plan B," literally manhandling them into place. Wrestling them into position by hand without being able to see around them and taking care not to bump into anything else in the payload bay was exhausting work.

The Manned Maneuvering Units, which enabled astronauts to move around and away from the orbiter untethered, were flown for the third and last time on *Discovery*, after being used on both *Challenger* missions earlier in the year. Altogether, six astronauts flew nine sorties and accumulated almost 10½ hours of flight time on two identical units. With insufficient demand for further use and lingering concerns about untethered EVAs, NASA took them out of service. One of the MMUs now "flies" near *Discovery* in the National Air and Space Museum's space hangar.

While outdoor work demanded much of the crew's attention, another first occurred quietly in the middeck. The first research project to develop organic crystals in space made good progress. Sponsored by the 3M Company, experiments were carried out in a reactor where, unaffected by gravity, chemical solutions formed into crystals of exceptional size and purity. Crystal growth in space emerged as an early and lasting interest shared by researchers in the electronics and pharmaceutical industries.

The first-ever retrieval and return of hardware from orbit demonstrated a shuttle service that NASA had promised its customers. Instead of being written off as complete losses, costly satellites could be plucked from space and salvaged. (Both Palapa and Westar were relaunched on rockets in 1990.) Scientific equipment or experiments could be flown, brought home, repaired or refined, and flown again. Such recycling fit into the shuttle era effort to make spaceflight more practical and economical.

Dale Gardner, flying the Manned Maneuvering Unit backpack, approaches Westar with a "stinger" capture device. When docked, he guided the satellite back to the shuttle.

Shuttle crews engaged in a friendly rivalry with signs like these for their phantom moving, delivery, and servicing "companies."

STS-51C: JANUARY 24–27, 1985

3 : 1 : 33
DAYS HRS MINS

212 M
185 NM
(345 KM)

49 | 28.5°

KENNEDY SPACE CENTER

0 : 0
HRS MINS

0

AN ALL-MILITARY SHUTTLE MISSION CREW

15TH SPACE SHUTTLE MISSION

COMMANDER

- Thomas K. (Ken) Mattingly II, USN, his 2nd shuttle flight; also flew on Apollo 16

PILOT

- Loren J. Shriver, USAF, his 1st of 3 flights, 2 on *Discovery*

MISSION SPECIALISTS

- Ellison S. Onizuka, USAF, aerospace engineer, his 1st of 2 flights

- James F. Buchli, USMC, aeronautical engineer, his 1st of 4 flights, 3 on *Discovery*

PAYLOAD SPECIALIST

- Gary E. Payton, USAF, astronautical-aeronautical engineer, his only flight

Mission Cloaked in Secrecy

The first shuttle mission reserved for the Department of Defense was cloaked in secrecy by a virtual news blackout. In the interests of national security, NASA did not issue a press kit for journalists, did not announce the exact launch time or mission duration nor identify the payload, and did not start the countdown clocks and commentary until nine minutes before launch. NASA TV did not broadcast live from the Mission Control Center, and the landing was not announced until the last hours in orbit. The all-military crew gave no interviews before or after flight. Never before had a space mission flown by U.S. astronauts been classified as secret.

The Department of Defense cautioned news media to refrain from speculating about the nature of the secret payload. Yet enough information was already available in public sources that the *Washington Post*, Associated Press, and *New York Times* reported what they presumed the shuttle was carrying: an advanced spy satellite capable of eavesdropping on the Soviet Union. Upon release from the shuttle, it was powered to geosynchronous orbit by an Air Force Inertial Upper Stage (IUS) booster rocket. The primary payload was never publicly confirmed but was speculated to be a Magnum signals intelligence satellite.

The STS-51C mission made visible a relationship between NASA and the Department of Defense that had developed largely behind the scenes during the early Space Shuttle era. The U.S. Air Force, National Reconnaissance Office, and National Security Agency were the agents for Defense missions.

The master plan for the shuttle program was to operate *Columbia* and *Challenger* from the Florida launch site and station *Discovery* at a

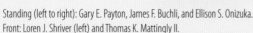

Apart from patriotic symbols and a star for each crewmember, the emblem yields no clues about the nature of this mission.

Standing (left to right): Gary E. Payton, James F. Buchli, and Ellison S. Onizuka. Front: Loren J. Shriver (left) and Thomas K. Mattingly II.

secure West Coast launch site at Vandenberg Air Force Base in California, where it would be dedicated to flying Department of Defense missions. *Discovery* was informally known as the "DOD shuttle" or the "Air Force shuttle." For a variety of reasons, none of this came to pass, and the shuttle launch complex at Vandenberg never served its purpose.

Nevertheless, the Department of Defense had priority to use the shuttle and already had reserved it for other secret missions. In the lingering Cold War political environment, the Soviet Union worried that the shuttle was a military vehicle. At home, the first classified shuttle mission stimulated public discussion about the militarization of space and secrecy in the U.S. space program.

Challenger had been scheduled to fly this mission but remained in the maintenance facility and *Columbia* was being refurbished, so *Discovery* substituted. The first launch attempt in late January 1985 was scrubbed when the temperature dropped to freezing and ice formed on the vehicle and pad, raising concerns about possible damage during launch. On the warmer next day, launch proceeded normally. The mission evidently went as planned, according to NASA's bland progress reports.

This launch was eerily similar to the catastrophic launch of *Challenger* the next January, also in freezing weather. Post-recovery examination of the solid rocket boosters that propelled *Discovery* toward space revealed an alarming amount of erosion, or hot gas "blow by," in rubber O-ring seals in the booster joints—the problem that led to burn-through and failure of *Challenger*'s 1986 launch. Ellison Onizuka made his first flight on *Discovery*'s classified mission, but he perished with six crewmates barely a minute into his next mission—on *Challenger*. This *Discovery* mission and crew came within a few millimeters of the same fate.

Discovery rose past the gantry on its third launch as well as the fifteenth shuttle launch, in January 1985. Erosion of seals in the solid rocket boosters, discovered later, foreshadowed the tragic loss of *Challenger* and its crew twelve months, and ten launches, later.

STS-51D: APRIL 12–19, 1985

6 :23 :55
DAYS HRS MINS

286 M
249 NM
(460 KM)

110 | 28.5°

3 : 6
HRS MINS

1

KENNEDY
SPACE CENTER

COMMANDER

- Karol J. Bobko, USAF, his 2nd of 3 flights

PILOT

- Donald E. Williams, USN, his 1st of 2 flights

MISSION SPECIALISTS

- Margaret Rhea Seddon, physician, her 1st of 3 flights, RMS operator
- Jeffrey A. Hoffman, astrophysicist, his 1st of 5 flights, 1 EVA
- S. David Griggs, research pilot, his only spaceflight, 1 EVA

PAYLOAD SPECIALISTS

- Charles D. Walker, McDonnell Douglas, test engineer, his 2nd of 3 flights, 2 on *Discovery*
- E. Jacob (Jake) Garn, U.S. Senate, his only spaceflight

Setting Records

By circumstance rather than plan, *Discovery* flew three missions in a row from November 1984 through April 1985. *Columbia* and *Challenger* did so when they were the only ships flying, but never as closely sequenced as STS-51A, -51C, and -51D. *Discovery* flew these back-to-back missions by taking *Challenger*'s place twice, first for the STS-51A flight and again this time. With *Atlantis* reporting for duty in October, 1985 became a banner year: Nine shuttle missions lifted off, and *Discovery* flew four of them to set an orbiter annual flight record that was never broken.

The STS-51D mission partially repeated -51A: deployment of two communications satellites—the third Canadian Telesat Anik and the third LEASAT (SYNCOM). On the first day, the crew sent the smaller Anik spinning out of the payload bay as usual and watched the boost stage ignite to send the satellite to its operational orbit. This release went flawlessly.

The next day, LEASAT rolled out as usual, but then nothing happened. Its antennas did not deploy and its rotation did not increase as expected. Evidently, the satellite had not switched on. Mission Control decided to try something new: the first unscheduled EVA, an unforeseen spacewalk, to activate the satellite manually.

The mission became an ingenious effort to avert failure by improvising a difficult rescue without prior training. As engineers and astronauts on the ground devised a solution, they sent instructions to the crew to use on-board materials to make something like a flyswatter and a lacrosse stick. They hoped that the makeshift tools would flip the activation switch on the stranded satellite. Spacewalking astronauts Jeff Hoffman and David Griggs suited up and went outside to attach the tools to the end of the orbiter's robotic arm.

The first official U.S. flag connects *Discovery* and spaceflight to the long national tradition of exploration.

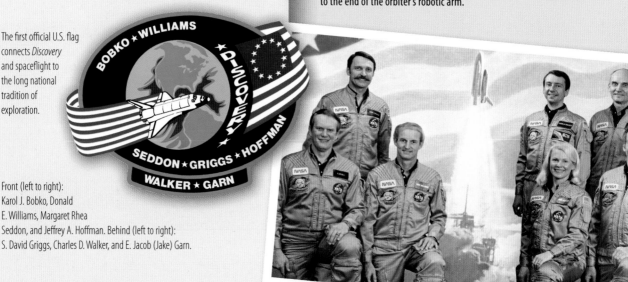

Front (left to right): Karol J. Bobko, Donald E. Williams, Margaret Rhea Seddon, and Jeffrey A. Hoffman. Behind (left to right): S. David Griggs, Charles D. Walker, and E. Jacob (Jake) Garn.

Commander "Bo" Bobko and pilot Don Williams then reoriented the orbiter at the correct angle near LEASAT for arm operator Rhea Seddon to attempt to "swat" the control switch. She made three solid contacts, but nothing happened; the switch itself had failed. The disappointed crew had to leave this satellite for another crew to rescue.

Discovery's crew included the first elected government official to fly in space, U.S. Senator Jake Garn of Utah. As chairman of the Senate appropriations subcommittee responsible for NASA's budget, he worked out an agreement to fly as an observer-payload specialist and participant in studies of space adaptation syndrome, or "space sickness." McDonnell Douglas payload specialist Charlie Walker flew again to operate the company's experiment on its sixth trip in space.

This mission had some other distinctions. An insect "crew" of three hundred houseflies flew while confined in a small habitat for a study of brain cell changes in microgravity. The flight literally ended with a bang when one of the landing gear tires blew out as *Discovery* rolled down the shuttle landing strip in Florida. Thereafter until the wheels and brakes were recertified, all landings occurred on the much longer desert lakebed runways at Edwards Air Force Base in California.

From start to finish, the STS-51D mission featured the resiliency of the shuttle program, crew, and entire mission team. *Discovery's* mission kept the flight schedule largely on course as the pace accelerated in the busiest shuttle year yet.

Above: The attempt to activate the satellite drew upon the skills of the entire crew. Spacewalkers Jeff Hoffman and David Griggs attached two tools handmade by Rhea Seddon and Jake Garn to the end of the remote manipulator arm.

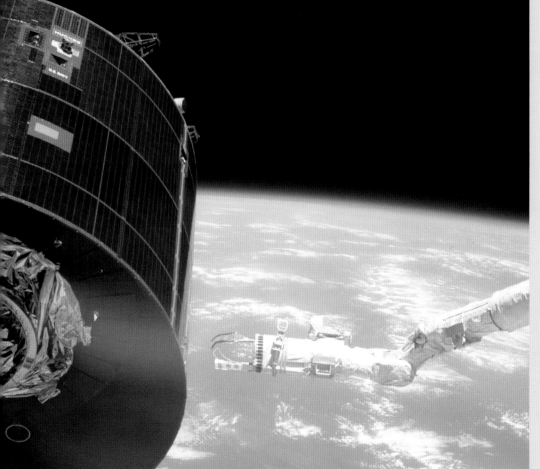

Left: The pilots flew the orbiter close enough to the satellite for Rhea Seddon to use the tool-equipped robotic arm to "flip the switch." The maneuvering worked, but the switch did not.

STS-51G: JUNE 17–24, 1985

7 : 1 : 38
DAYS   HRS   MINS

112 | 28.5°

240 M
209 NM
(386 KM)

0 : 0
HRS   MINS

0

EDWARDS
AIR FORCE BASE

5TH *Discovery* MISSION

COMMANDER

- Daniel C. Brandenstein, USN, his 2nd of 4 flights

PILOT

- John O. Creighton, USN, his 1st of 3 flights, 2 on *Discovery*

MISSION SPECIALISTS

- Shannon W. Lucid, biochemist, her 1st of 4 flights plus roundtrip rides for a 188-day stay on Mir

- John M. Fabian, aeronautical-astronautical engineer, his last of 2 flights

- Steven R. Nagel, engineer, his 1st of 4 flights

PAYLOAD SPECIALISTS

- Patrick Baudry, France, aeronautical engineer and pilot, his only flight

- Sultan Salman Abdulaziz Al-Saud, Saudi Arabia, government official and pilot, his only flight

The Wright Flyer and Space Shuttle signify the advance from aviation to spaceflight.

Front (left to right): Daniel C. Brandenstein and John O. Creighton.
Back (left to right): Shannon W. Lucid, Steven R. Nagel, John M. Fabian, Sultan Salman Abdulaziz Al-Saud, and Patrick Baudry.

International Flair

Discovery's fifth mission was its first to include crewmembers from other nations. *Columbia* and *Challenger* had already flown with payload specialists from Germany and Canada. Now a member of the royal family of Saudi Arabia and an astronaut from France boarded *Discovery* on the first spaceflight to include citizens of three nations. Both Sultan Salman Al-Saud and Patrick Baudry were accomplished pilots, but on the shuttle they had research duties.

Once again, *Discovery*'s crew deployed three communications satellites: Mexico's first MORELOS; an ARABSAT for the Arab League; and another TELSTAR for AT&T. All reached their planned orbits and began operating as expected. Using the remote manipulator arm, the crew for the first time deployed and later retrieved a small SPARTAN scientific satellite loaded with astronomical instruments that flew in formation with *Discovery* but about 100 miles away. Upon return to Earth, it was reconfigured and flown again on several future missions.

A variety of scientific payloads consumed much of the crew's attention. They participated in two French biomedical experiments and used a research furnace in the middeck crew cabin for an investigation of magnetic materials. They also remotely activated six Get Away Special experiments (three from Germany) stowed in canisters in the payload bay, and they carried out a program of Earth photography from the flight deck windows.

A different kind of experiment captured the most media attention: a trial of a laser system with possible utility for the "Star Wars" strategic

The reusable SPARTAN scientific satellite was deployed for free-flight and retrieved by the orbiter's robotic arm, as seen here. A mission specialist operated the arm from the aft flight deck.

defense initiative championed by President Ronald Reagan and the Department of Defense. Officially called the High-Precision Tracking Experiment, the trial involved shining a powerful laser at the shuttle passing overhead. An 8-inch mirror attached to the side of the shuttle reflected the beam to the ground station, enabling the laser to stay "locked on" to the "target." Researchers sought the test to evaluate whether ground-based lasers might be feasible for tracking or destroying hostile missiles in space.

This weeklong mission ended on a dry lakebed in the Mojave Desert, the first of several required landings in California until a recurrent brake damage problem was resolved. With more space to roll out after touchdown and less threat of crosswinds than at the Florida runway, the desert site allowed for less strenuous braking.

Like previous satellite deployment and Spacelab missions, STS-51G demonstrated opportunities for the international community to place people and payloads on the Space Shuttle. NASA's spokesman at the landing called this one of the most successful missions to date, as measured by the almost flawless performance of the orbiter and meeting of all mission objectives. *Discovery* also passed a crew milestone: With Shannon Lucid on board, all six of the first U.S. women astronauts had now flown in space. Forthright as usual, she expressed her view of mission success: "We had a job to do, and we went and did it."

Leaving crew quarters to go to the launch pad, Patrick Baudry sported a beret as the first French citizen on a shuttle mission. Sultan Al-Saud, the first Arab in space and at age twenty-eight the youngest person ever to fly on the shuttle, is third from the right in this photo of *Discovery*'s first international crew.

7 : 2 : 17
DAYS HRS MINS

11 : 46
HRS MINS

2

278 M
242 NM
(447 KM)

112 | 28.5°

EDWARDS
AIR FORCE BASE

20TH SPACE SHUTTLE MISSION

COMMANDER

- Joe H. Engle, USAF, his 2nd of 2 orbital flights after Approach and Landing Test flights on *Enterprise*

PILOT

- Richard O. Covey, USAF, his 1st of 4 flights, 2 on *Discovery*

MISSION SPECIALISTS

- James D. A. van Hoften, hydraulic engineer, his last of 2 flights, 2 EVAs

- John M. (Mike) Lounge, astrogeophysicist, his 1st of 3 flights, 2 on *Discovery*, RMS operator

- William F. Fisher, physician, his only flight, 2 EVAs

Release, Retrieve, and Repair

In late August, *Discovery* blazed into space on its fourth 1985 mission, setting an orbiter annual flight record never broken during the Space Shuttle era. Once again, three communications satellites filled its payload bay. The five-man crew relished their additional assignment: to retrieve and repair the disabled LEASAT (SYNCOM) deployed by a *Discovery* crew in April. This twentieth shuttle mission would showcase the orbiter as both a space truck and a service station.

The crew first attended to the three new satellites, planning to release one a day at the beginning of the mission. First up: ASC-1 for the American Satellite Company to serve businesses and government agencies; then AUSSAT-1, the first of three planned communications satellites for Australia; and finally another LEASAT (SYNCOM) for the U.S. Navy. These deployments proceeded like a familiar routine. In fact, the crew deftly adjusted their plans and deployed AUSSAT on the same day as the ASC when its protective enclosure misbehaved.

The next task was more challenging for the entire crew. After rendezvous with the dormant satellite, "Ox" van Hoften and Bill Fisher suited up and went out to steady the almost 8-ton, slowly spinning cylinder and assist RMS arm operator Mike Lounge in guiding it into the payload bay. This became more complicated when the crane-like arm malfunctioned. The one-EVA job expanded into two EVAs, during which the spacewalking team rewired the timer for the boost motor.

That work went well and raised confidence that the repair would succeed. Ground controllers revived the satellite, and later the engine

Patriotic icons and colors set the theme for this mission.

Gathered in the airlock of the crew compartment trainer at Johnson Space Center, clockwise from mission emblem: William F. Fisher, Richard O. Covey, Joe H. Engle, James D. A. van Hoften, and John M. (Mike) Lounge in the center.

fired as it should, sending LEASAT (SYNCOM) to its proper orbit, to the relief of its owner, lessee, and insurance company. Meanwhile, *Discovery*'s crew once again demonstrated expertise in fixing something that had not been designed for repair, an experience that would prove valuable on future missions.

The mission wrapped up a day ahead of schedule, and *Discovery* returned to California. *Atlantis*, the newest addition to the shuttle fleet, was on the launch pad in Florida for its first mission in early October, and then three more flights of *Challenger*, *Atlantis*, and *Columbia* completed the year for a total of nine missions. It was the busiest year yet for the shuttle, almost meeting the target of a mission every month. The next year looked to be even more ambitious.

Events in the background that held no particular importance for *Discovery* at the time soon became significant. During the summer of 1985

while *Discovery* flew twice, the field of candidates for the Teacher-in-Space shuttle flight narrowed, and Christa McAuliffe was selected. Several 1985 post-mission reports noted evidence of damaged seals between solid rocket booster segments. Those unrelated events came together tragically in the second mission of 1986, when the STS-51L *Challenger* mission ended fatally during ascent. NASA grounded the shuttle program to investigate and remedy the cause of the accident, and flights did not resume until September 1988, when *Discovery* led the way back into space.

Above right: James D. "Ox" van Hoften, mounted on *Discovery*'s robotic arm, grasped the LEASAT (SYNCOM) satellite, first to slow its spin for retrieval and later to give it some spin for redeployment.

Right: Another successful deployment of a communications satellite attached to a solid-rocket boost motor marked this mission, this time for Australia.

4 : 1 : 0
DAYS   HRS   MINS

205 M
178 NM
(330 KM)

64 | 28.5°

0 : 0
HRS   MINS

0

EDWARDS
AIR FORCE BASE

COMMANDER

- Frederick H. (Rick) Hauck, USN, his last of 3 flights, 2 on *Discovery*

PILOT

- Richard O. (Dick) Covey, USAF, his 2nd of 4 flights, 2 on *Discovery*

MISSION SPECIALISTS

- John M. (Mike) Lounge, astrogeophysicist, his 2nd of 3 flights, 2 on *Discovery*

- George D. "Pinky" Nelson, astronomer, his last of 3 flights

- David C. Hilmers, USMC, electrical engineer, his 2nd of 4 flights, 2 on *Discovery*

Symbolism dominated this emblem. Seven stars commemorated the *Challenger* crew as spaceflight resumed on a new day in NASA's legacy. (The red vector is from the agency's original logo.)

This *Discovery* crew first wore the new launch-entry pressure suits. Seated: Richard O. Covey (left) and Frederick H. Hauck. Standing (left to right): John M. (Mike) Lounge, David C. Hilmers, and George D. Nelson.

Return to Flight

Discovery's next trip to the launch pad occurred almost three years later, on Independence Day, 1988. When OV-103 launched in late September, its ostensible mission was to deliver another communications satellite, but its real purpose was to restore confidence in the Space Shuttle, NASA, and the U.S. human spaceflight program. *Discovery* put America back in space.

Managers had selected OV-103 for the return-to-flight mission shortly after the 1986 STS-51L *Challenger* tragedy. *Atlantis* was already configured for its next flight and *Columbia* had just entered servicing after its latest return, but *Discovery* was then idle. In fact, *Discovery* was waiting to be moved to California later in 1986 and prepared to fly the first shuttle mission from Vandenberg Air Force Base. Its readiness for the return-to-flight assignment became the pivot point in *Discovery*'s career. The West Coast launch site was canceled for unrelated reasons, and *Discovery* stayed in Florida, eventually becoming the most flown orbiter in the fleet.

In preparation for its next flight, technicians gave *Discovery* a thorough tune-up and made about two hundred changes to the orbiter, approximately one hundred of them required as outcomes of the accident investigation. Major improvements were made to the brakes, all engines and thrusters, and some areas of thermal tiles and blankets. Simultaneously, the solid rocket booster joints were redesigned and other changes were made elsewhere in the boosters and external tank. The long pause between shuttle flights occasioned many technical and organizational changes for improved safety.

The STS-26 crew also adopted new safety measures, most noticeably a bright orange partial pressure suit. After the first four test flight missions, shuttle astronauts no longer wore pressure suits for launch and reentry; ordinary fabric flight suits were deemed sufficient.

After the *Challenger* accident, crews donned pressure suits for these two critical phases of spaceflight. The new Launch-Entry Suit (LES), familiarly known as the "pumpkin suit," gave the crew some protection for loss of cabin pressure, fire, or emergency bailout. It included a parachute pack and survival gear.

Two crew escape aids were installed in the middeck: a telescoping pole and an inflatable slide. In an emergency during ascent or descent, the crew could blow out the hatch, release the long pole to extend through the open hatch, attach their harness to a sliding hook, and bail out one at a time to descend by parachute. In an emergency landing, they could blow the hatch and trigger the slide for a rapid exit. *Discovery* was the first orbiter to fly with these devices; fortunately, they never had to be used on any flight.

The mission's primary objective was to deploy a second Tracking and Data Relay Satellite (TDRS); the first was launched in 1983. It also had been the objective of *Challenger*'s last mission, but that satellite was destroyed in the accident, so *Discovery* actually carried the third one. With two of these large communications satellites in service, NASA would have nearly continuous contact with the shuttle and other scientific satellites in low Earth orbit—such as LANDSAT and the upcoming Hubble Space Telescope—as an alternative to partial coverage through a network of ground stations.

Discovery's crew deployed the TDRS six hours after launch by commanding its support stand to tilt it upward; then explosive bolts broke and a spring-loaded mechanism pushed the satellite out of the payload bay. Its powerful upper stage fired an hour later to propel TDRS about 22,300 miles (35,400 km) to geosynchronous orbit.

For the rest of the mission, the crew focused on eleven materials-processing and biotechnology experiments in the middeck. Companies providing experiments included 3M, DuPont, Merck, Upjohn, and others, plus university researchers. The crew also followed an Earth observations plan and took about two thousand photographs.

Starting with STS-26, NASA abandoned the alphabetical mission designations and reverted to the simpler numeric scheme. All subsequent missions had sequential numbers, although they did not always fly in order.

The STS-26 crew was the first composed entirely of experienced spaceflight veterans. Commander Rick Hauck remarked that they felt the responsibility of "getting us back in the manned spaceflight business." In the words of Mission Control upon landing, it was a fine ending to a new beginning in space.

Amid the euphoria of *Discovery*'s successful return to flight, inspectors found something ominous. An area of tiles under one wing was badly damaged by a debris strike during launch and heat erosion during reentry. A flaw in the orbiter's protective shield could leave it vulnerable during descent. Although no one could recognize it as such at the time, this damage foreshadowed the next shuttle tragedy fifteen years later.

Discovery heads through the clouds on the 1988 return-to-flight mission.

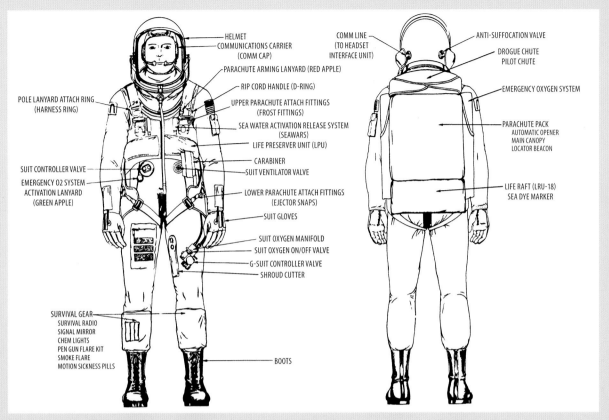

Survival gear packed with the Launch-Entry Suit and parachute included a life preserver, life raft, and several smaller items.

The labels on the diagram, from top left clockwise:

HELMET
COMMUNICATIONS CARRIER (COMM CAP)
PARACHUTE ARMING LANYARD (RED APPLE)
RIP CORD HANDLE (D-RING)
UPPER PARACHUTE ATTACH FITTINGS (FROST FITTINGS)
SEA WATER ACTIVATION RELEASE SYSTEM (SEAWARS)
LIFE PRESERVER UNIT (LPU)
CARABINER
SUIT VENTILATOR VALVE
LOWER PARACHUTE ATTACH FITTINGS (EJECTOR SNAPS)
SUIT GLOVES
SUIT OXYGEN MANIFOLD
SUIT OXYGEN ON/OFF VALVE
G-SUIT CONTROLLER VALVE
SHROUD CUTTER
BOOTS

POLE LANYARD ATTACH RING (HARNESS RING)
SUIT CONTROLLER VALVE
EMERGENCY O2 SYSTEM ACTIVATION LANYARD (GREEN APPLE)
SURVIVAL GEAR
SURVIVAL RADIO
SIGNAL MIRROR
CHEM LIGHTS
PEN GUN FLARE KIT
SMOKE FLARE
MOTION SICKNESS PILLS

COMM LINE (TO HEADSET INTERFACE UNIT)
ANTI-SUFFOCATION VALVE
DROGUE CHUTE PILOT CHUTE
EMERGENCY OXYGEN SYSTEM
PARACHUTE PACK
AUTOMATIC OPENER
MAIN CANOPY
LOCATOR BEACON
LIFE RAFT (LRU-18) SEA DYE MARKER

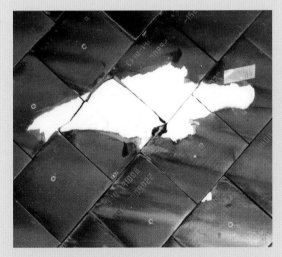

On the return-to-flight mission, *Discovery* suffered the worst damage yet to protective tiles. This area was struck by debris from one of the solid rocket boosters during launch, leaving the orbiter vulnerable during reentry.

Opposite: The Tracking and Data Relay Satellite attached to its 17-foot-long (5.2 meters) two-stage solid rocket weighed about 20 tons. Upon reaching the proper orbit, its solar arrays and gold antennas extended.

In a show of national pride, Vice President George H. W. Bush welcomed the *Discovery* crew home.

STS-29: MARCH 13–18, 1989

4:23:38 DAYS HRS MINS

0:0 HRS MINS — 0

205 M
178 NM
(330 KM)

80 | 28.5°

EDWARDS
AIR FORCE BASE

COMMANDER

- Michael L. Coats, USN, his 2nd of 3 flights, all on *Discovery*

PILOT

- John E. Blaha, USAF, his 1st of 6 shuttle flights, 2 on *Discovery*, and a stay on Mir

MISSION SPECIALISTS

- James F. Buchli, USMC, aeronautical engineer, his 3rd of 4 flights, 3 on *Discovery*

- Robert C. Springer, USMC, operations research and systems analysis, his 1st of 2 flights

- James P. Bagian, medical doctor and engineer, his 1st of 2 flights

This design suggests the powerful forward thrust of the shuttle and the space program; seven stars commemorate the *Challenger* crew.

Seated: John E. Blaha (left) and Michael L. Coats. Standing (left to right): James P. Bagian, Robert C. Springer, and James F. Buchli.

Another Satellite for NASA

Five months later, *Discovery* opened 1989 by launching on a mission similar to its last. Again it carried a Tracking and Data Relay Satellite (TDRS) for NASA and a variety of scientific experiments for commercial and university customers. The payload bay also held a technology experiment for NASA's own researchers.

This TDRS was the fourth to be launched but the third to reach orbit, the second one having been destroyed with *Challenger*. It replaced the first TDRS, operational since 1983, which was repositioned as an in-orbit spare. Two operational TDRS high above the equator, one over the Atlantic Ocean and one over the Pacific, gave NASA almost continuous contact with the shuttle and scientific satellites in low Earth orbit. Both were deployed from *Discovery* and served more than twenty years.

After delivering the satellite on their first day in space, the crew kept a busy research schedule. In the middeck, a protein crystal growth facility flew again, hosting sixty experiments from university, government, and industry research teams. Another experiment studied cell division and genetics in the roots of plants sprouting in microgravity. Thirty-two fertilized eggs and four rats went on the flight in two student-developed experiments that examined chicken embryo development and bone healing in weightlessness. In the payload bay, a heat pipe experiment tested a new radiator technology being considered for future space station use. The crew activated and monitored all experiments for scientists on the ground, who followed their progress intently.

Discovery also carried an IMAX® camera, as it had on its first mission. Crewmembers trained with IMAX® experts to get the best, most desirable shots. Earth views from these missions became part of *Blue Planet*, the second IMAX® feature film shot in space by astronauts.

As STS-29 ended, the Space Shuttle program was off to a good start in a five-mission year.

Pilot John E. Blaha also served as
cinematographer, using the 70 mm
IMAX® camera to shoot Earth views
through the windows.

After a two-hour delay until morning fog
lifted, *Discovery* made the first launch
of 1989.

5 : 0 : 6
DAYS   HRS   MINS

79 | 28.5°

348 M
302 NM
(559 KM)

0 : 0
HRS   MINS

0

EDWARDS
AIR FORCE BASE

COMMANDER

• Frederick D. Gregory, USAF, his 2nd of 3 flights

PILOT

• John E. Blaha, USAF, his 2nd of 6 flights, 2 on *Discovery*, and a stay on Mir

MISSION SPECIALISTS

• Manley L. Carter Jr., USN, medical doctor, his only flight

• F. Story Musgrave, USMC, medical doctor, his 3rd of 6 flights

• Kathryn C. Thornton, physicist, her 1st of 4 flights

The symbol for this national security mission was a falcon with wings suggesting the flag's stripes against a field of stars. The single gold star on the border of crew names represented the original pilot, S. David Griggs.

Arranged clockwise around Frederick D. Gregory (center): Kathryn C. Thornton and Manley L. "Sonny" Carter Jr. (left); John E. Blaha, and F. Story Musgrave (right).

Mystery and History

Discovery flew the last mission of 1989, its second classified mission for the Department of Defense. This was actually the fifth in a series of ten mysterious national security missions, all but one of which were flown from 1984 to 1992 by *Discovery* or *Atlantis*. All members of this crew had military experience.

As before, NASA did not issue a press kit about the payload and crew activities nor provide commentary during the flight, and the crew made no public appearances related to this mission. News reports included speculation that the payload was the same as that on STS-51C, the first classified mission. It was believed to be a Magnum electronic intelligence satellite operated by the National Reconnaissance Office for listening to official communications in and between Communist nations. After the mission, the crew reported on their unclassified work doing biomedical and technology experiments and Earth photography.

Frederick D. Gregory made history as the first African American commander of a space mission. He had already served as first African American space pilot on a 1985 shuttle-Spacelab flight and would command another Department of Defense mission in 1991. Gregory later held senior leadership positions at NASA Headquarters.

Tragedy touched the crew several months before flight when the scheduled pilot, S. David Griggs, died off duty in the crash of a private aircraft. STS-33 would have been his second flight on *Discovery*. John Blaha, pilot on the previous *Discovery* mission, stepped into the pilot role. Another crewmember, "Sonny" Carter, died in a commercial airplane crash in 1991. The other members of this crew continued to fly on missions in the 1990s.

Discovery made the first night launch after the *Challenger* accident, only the third launch in darkness thus far. Even the television reporters called it spectacular.

STS-31: APRIL 24–29, 1990

5 : 1 : 16
DAYS   HRS   MINS

80 | 28.5°

383 M
333 NM
(616 KM)

0 : 0
HRS   MINS

0

EDWARDS
AIR FORCE BASE

10TH *DISCOVERY* MISSION

35TH SPACE SHUTTLE MISSION

COMMANDER
- Loren J. Shriver, USAF, his 2nd of 3 flights, 2 on *Discovery*

PILOT
- Charles F. Bolden Jr., USMC, his 2nd of 4 flights, 2 on *Discovery*

MISSION SPECIALISTS
- Steven A. Hawley, astronomer and astrophysicist, his 3rd of 5 flights, 3 on *Discovery*; RMS operator
- Bruce McCandless II, USN, electrical engineer, his last of 2 flights
- Kathryn D. Sullivan, USN, geologist and oceanographer, her 2nd of 3 flights

Hubble Space Telescope Delivery

Discovery's first flight of the new decade, its tenth and the shuttle program's thirty-fifth mission, created a media stir. The crew would deliver into orbit the long-anticipated Hubble Space Telescope, the first in a family of four "Great Observatories" designed to explore the universe across the electromagnetic spectrum.

Originally scheduled to be launched in 1986 on *Atlantis*, the telescope was postponed first by the *Challenger* accident and then by reorganized priorities when shuttle flights resumed in 1988. Nine other missions flew before *Discovery* stood on the launch pad with Hubble in the payload bay. *Columbia* was being readied on the other pad, only the second time that both launch pads were occupied simultaneously; the first had occurred in January 1986 with *Columbia* and *Challenger* both poised for what was to have been the shuttle's busiest year.

The Hubble Space Telescope was NASA's premier new program for astronomy and astrophysics, advertised as surpassing the best observatories on Earth at the time, designed to see deeper in space and farther back in time toward the dawn of the universe. It held promise to answer some of the most profound questions about the universe.

To meet such high expectations, *Discovery*'s delivery crew was well prepared for any foreseeable problem. Their job was to lift the telescope out of the payload bay, oversee the extension of its solar arrays and antennas, release it into space, and make sure the aperture door opened—all actions to be monitored from the flight deck. However, in case any appendage failed to extend or the aperture door remained closed, Bruce McCandless and Kathy Sullivan had trained for years to go outside and manually configure it.

The Hubble Space Telescope points toward the universe. The spiral galaxies and the red swath behind the shuttle acknowledge Sir Edwin Hubble's research into the nature of galaxies, red shift, and the expansion of the universe.

From left to right: Charles F. Bolden, Jr., Steven A. Hawley, Loren J. Shriver, Bruce McCandless II, and Kathryn D. Sullivan.

Grasped by the robotic arm and seen through an overhead window, the Hubble Space Telescope has one solar array wing and both antennas extended. After the second solar array unfurled, Steven A. Hawley guided the arm to release the telescope.

As deployment began on the second day in orbit, the two spacewalkers prepared in advance for that contingency by prebreathing oxygen, organizing their tools, and partially suiting up to save time in case an extravehicular activity proved necessary. Steve Hawley operated the remote manipulator arm to lift the telescope and rotate it into proper position, while Loren Shriver and Charlie Bolden did spotting and photography. When the solar panels unfurled and the antennas extended on command, there was no need for spacewalking assistance, and Hawley released the telescope. Three days later, the aperture door was commanded open to receive first light, and again, the operation succeeded without the need for EVA help.

Discovery reached the highest altitude then to date, placing the Hubble Space Telescope in a 383-mile (616-kilometer) orbit. The crew used two IMAX® cameras—one mounted in the payload bay and the other handled in the crew cabin—to film the deployment sequence and also the broader views of Earth from this high vantage point. They also attended to eight experiments, several of them on repeat flights for additional experience in protein crystal growth and other matters. *Discovery* returned home from a well-executed mission and smoothly rolled to a stop using new carbon brakes for the first time.

During the next several weeks of remotely controlled telescope activation and checkout, engineering and science teams discovered a focus flaw in the primary mirror that would spark the first servicing mission three years later. *Endeavour* flew that mission, but *Discovery* flew the second and third servicing missions in 1997 and 1999; it is the only orbiter to engage with the Hubble Space Telescope three times.

Suited up and ready: Kathryn D. Sullivan and Bruce McCandless II were prepared to handle any problems with telescope deployment. Both had already made EVA history, she as the first U.S. female spacewalker and he as the first human satellite, free-flying with a propulsion backpack.

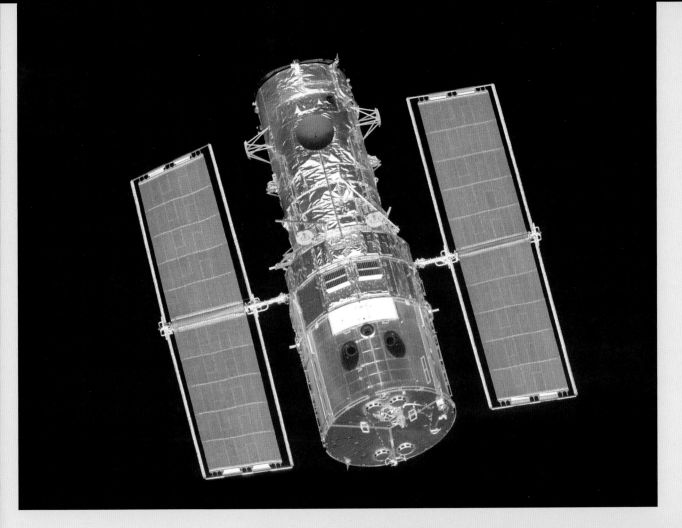

Above: Other shuttle crews had this view of the telescope as they approached it. The aperture door was already closed to protect the mirrors during servicing missions.

Right: After the telescope's release into space, the aperture door opened to receive light from the farthest, faintest, most mysterious features of the universe.

Opposite: The crew and cameras had an expanded field of view from an altitude up to 225 miles higher than prior shuttle missions. Here, much of Egypt, the Red Sea, Saudi Arabia, and areas of the Mediterranean Sea, Israel, and Jordan are visible.

4 : 2 : 10
DAYS HRS MINS

66 | 28.5°

184 M
160 NM
(296 KM)

0 : 0
HRS MINS

0

EDWARDS
AIR FORCE BASE

COMMANDER

- Richard N. Richards, USN, his 2nd of 4 flights, 2 on *Discovery*

PILOT

- Robert D. Cabana, USMC, his 1st of 4 flights, 2 on *Discovery*

MISSION SPECIALISTS

- Bruce E. Melnick, USCG, aeronautical engineer, his 1st of 2 flights

- William M. Shepherd, USN, mechanical engineer, his 2nd of 4 shuttle flights and an ISS expedition

- Thomas D. Akers, USAF, mathematician, his 1st of 4 flights

The red streak depicts the journey of the Ulysses probe from the shuttle, around Jupiter, and on to orbit the sun.

Front: Robert D. Cabana (left) and Richard N. Richards. Back (left to right): Bruce E. Melnick, Thomas D. Akers, and William M. Shepherd.

Solar Polar Expedition

Later in 1990, a *Discovery* crew released Ulysses, the first spacecraft to fly past the polar regions of the sun. Ulysses was the third interplanetary probe to start its journey from the shuttle; Magellan and Galileo departed from *Atlantis* in 1989. Almost four years later, Ulysses began its initial transit of the sun.

The crew deployed Ulysses on the first day in space. Like shuttle-deployed communications satellites, it was attached to rocket stages for a boost out of low Earth orbit. Both an Inertial Upper Stage (IUS) and a Payload Assist Module (PAM) propelled Ulysses into interplanetary space.

Ulysses traveled first to Jupiter for an extraordinary change in direction. The gravity of the giant planet bent the small spacecraft's path out of the plane of the planets into a perpendicular orbit that took it over the north and south poles of the sun. Its scientific instruments had the first good look at those regions and also at the solar magnetic fields and solar wind in the space environment around the sun. A joint project of NASA and the European Space Agency, Ulysses made six passes over the solar poles in its eighteen years of service.

The next four days in orbit were devoted to scientific and engineering work. Two technology experiments rode in the payload bay, and the middeck carried a group of materials processing, plant growth, physiology, radiation monitoring, and combustion experiments. Some of the equipment from NASA, university, and commercial researchers had flown before. Neither the small laboratory rats nor flames escaped from their enclosed chambers.

This mission, like many others, demonstrated the versatility of the Space Shuttle. Using it as a launch pad in low Earth orbit had an advantage over launching from the ground: Interplanetary spacecraft could be propelled by smaller, less expensive rockets. At the same time, the shuttle supported research in different fields, and scientists could fly experiments more than once to expand on previous results. These practices signaled the shuttle program's efforts to make access to space more economical.

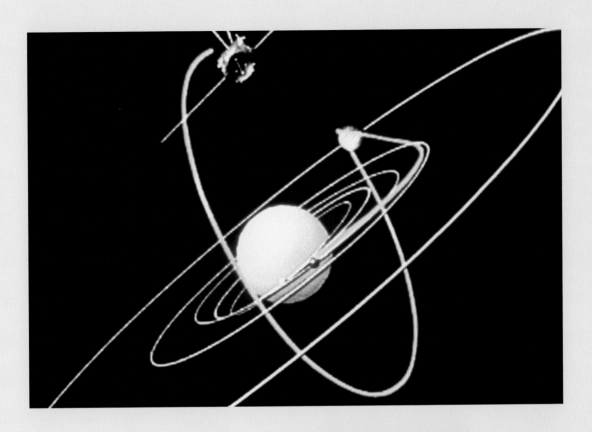

Above: Ulysses's flight path made a sharp turn at Jupiter, putting the spacecraft on course to pass by the sun from pole to pole.

Left: Bruce E. Melnick, the first astronaut from the U.S. Coast Guard, followed custom by flying his service banner on board.

8 : 7 : 22
DAYS HRS MINS

134 | 57°

161 M
140 NM
(259 KM)

0 : 0
HRS MINS

0

KENNEDY SPACE CENTER

40TH SPACE SHUTTLE MISSION

COMMANDER

- Michael L. Coats, USN, his last of 3 flights, all on *Discovery*

PILOT

- L. Blaine Hammond Jr., USAF, his 1st of 2 flights, both on *Discovery*

MISSION SPECIALISTS

- Gregory J. Harbaugh, engineer, his 1st of 4 flights, 2 on *Discovery*

- Donald R. McMonagle, USAF, mechanical engineer, his 1st of 3 flights

- Guion S. (Guy) Bluford Jr., USAF, aerospace engineer, his 3rd of 4 flights, 2 on *Discovery*

- Charles Lacy Veach, USAF, engineer, his 1st of 2 flights

- Richard J. Hieb, aerospace engineer, his 1st of 3 flights

Unclassified Defense Research

Having flown two classified Department of Defense (DOD) missions, *Discovery* now flew the first unclassified one open to public scrutiny. It was the eighth dedicated DOD mission. From this point on, national security shuttle missions were no longer secret, even if specific payloads were classified. Finally, NASA could be open about military activity on the Space Shuttle, and news media could offer real information rather than speculation.

For its longest mission to date, eight days, *Discovery* was loaded with research instruments to observe and probe the space environment. Some devices passively recorded natural phenomena, such as auroras and atmospheric airglow. Others actively perturbed the space environment by releasing certain chemicals and watching what happened. Sensors and low-light television cameras recorded these phenomena for detailed study.

The crew worked in two shifts to conduct an intensive research program sponsored by the U.S. Air Force (USAF) and the Strategic Defense Initiative Organization (SDIO). Most of the experiments focused on testing anti-missile sensors or distinguishing propellant gases from natural light effects in space. The broad goal was learning how to identify the optical and chemical signatures of missile engines firing in space in order to develop effective anti-missile defense technologies.

The USAF payload instruments observed sources of light—both natural and artificial—across the spectrum from infrared to x-ray wavelengths. Located on a platform straddling the payload bay and operated by the crew from the aft flight deck, these devices gathered data from the aurora, atmosphere, astrophysical sources, and natural background radiation. Better understanding of those sources would aid development of sensors to detect and track emissions from missiles and spacecraft.

In this arrowhead shape aimed high, the seven-star constellation Aquila (eagle) represents the crew and the electromagnetic spectrum represents the mission's focus on observations of the space environment.

HARBAUGH HIEB VEACH BLUFORD MCMONAGLE COATS HAMMOND

From left to right: Charles Lacy Veach, Donald R. McMonagle, Gregory J. Harbaugh, Michael L. Coats (center), L. Blaine Hammond Jr., Richard J. Hieb, and Guion (Guy) S. Bluford Jr.

The SDIO research relied on instruments in the payload bay and on a Shuttle Pallet Satellite (SPAS) that was deployed to operate a few miles from the shuttle and then was retrieved. Together, they observed firings of the orbiter's maneuvering thrusters and releases of typical rocket propellant liquids and gases. As these chemicals erupted from small canisters mounted in the payload bay or ejected from the vehicle, they formed glowing clouds that mimicked rocket exhaust plumes. Sensors on the shuttle and SPAS, at an air force ground station, and on an aircraft were trained on these clouds. The experiments yielded information for better detection and identification of chemicals that might be associated with missiles.

Some news reporters called the mission a smartly choreographed ballet or a slow-motion acrobatic program because the orbiter had to make more than sixty complicated maneuvers to be in the right position for each experiment. To point instruments in the payload bay toward the polar auroras or the airglow seen along Earth's horizon, the entire orbiter had to turn. Chemical releases and engine firings also required special maneuvers to ensure they were visible to instruments on the SPAS. Timing of the active experiments in coordination with all observing sites had to be exact. Piloting the orbiter to rendezvous with and retrieve the satellite required extra precision and skillful control.

Other air force payloads observed the fringes of Earth's atmosphere, cloud formations, and the luminous glow of ionized atomic oxygen that outlined parts of the orbiter as it moved through space. In addition, one classified USAF payload was deployed from a canister in the payload bay on the last day in orbit.

By most measures, this was the most complex shuttle mission to date, and it gave OV-103 a few more firsts. It was *Discovery*'s first mission with crew operations around the clock, and it was the tallest crew yet with an average height above six feet. *Discovery* flew its first high-inclination trajectory, 57° from the equator, traveling over more of the Earth than on a typical 28° path, to get better views of the auroras and atmospheric airglow and perhaps as required for the classified payload. *Discovery* flew with five improved general-purpose computers, the brain that controlled the orbiter's systems. And *Discovery* landed in Florida again, its first landing at home since the braking problems made California the preferred landing site.

Discovery's pilot, Blaine Hammond, proudly summed up STS-39 as "the grand slam of all shuttle missions."

Discovery participated actively in experiments to analyze the chemical signature of its thruster firings in an effort to develop more effective missile detection sensors.

Five new computers that controlled orbiter systems were half the size but operated faster and had more memory than the originals.

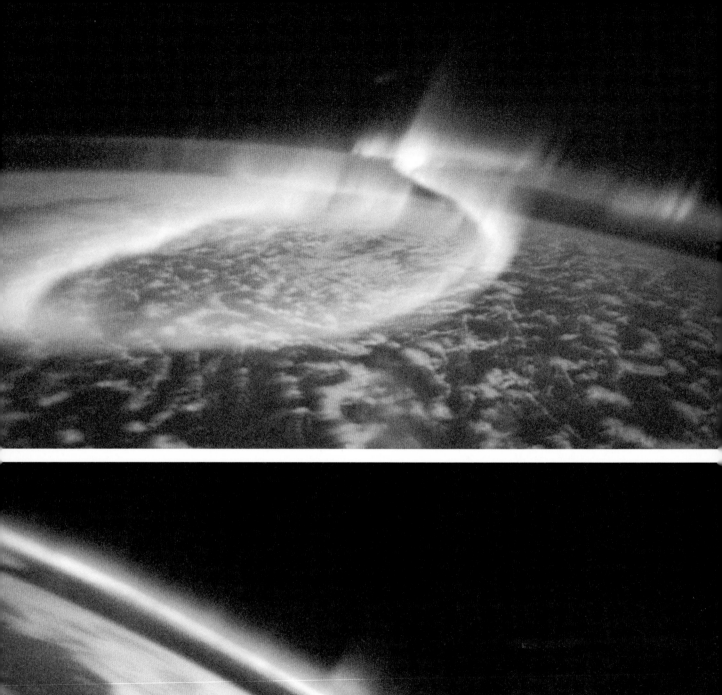

Above: The crew used the orbiter's mechanical arm to move the Shuttle Pallet Satellite (SPAS) out of the payload bay for its observing program and later retrieved it. SPAS flew about 6 miles ahead of and above *Discovery*.

Opposite, top: The southern lights, or *aurora australis*, curve and fold like sheer curtains above the Earth.

Opposite, bottom: A green aurora and the glowing atmosphere are vivid in this night pass over the Southern Hemisphere.

STS-48: SEPTEMBER 12–18, 1991

5 : 8 : 27
DAYS HRS MINS

81 | 57°

356 M
309 NM
(572 KM)

0 : 0
HRS MINS

0

EDWARDS
AIR FORCE BASE

COMMANDER

- John O. Creighton, USN, his last of 3 flights, 2 on *Discovery*

PILOT

- Kenneth S. Reightler Jr., USN, his 1st of 2 flights, both on *Discovery*

MISSION SPECIALISTS

- Charles D. (Sam) Gemar, USA, engineer, his 2nd of 3 flights

- James F. Buchli, USMC, aeronautical engineer, his last of 4 flights, 3 on *Discovery*

- Mark N. Brown, USAF, aeronautical engineer, his last of 2 flights

The triangular shape symbolizes the relationship of chemistry, physics, and energy in the upper atmosphere. Colors of the spectrum link the satellite and the atmosphere, and the stars of the northern sky as seen in fall and winter represent the mission's timing.

From left to right around John O. Creighton (center): Mark N. Brown, Charles D. (Sam) Gemar, James F. Buchli, and Kenneth S. Reightler Jr.

Mission to Planet Earth

Once again, *Discovery* hauled a scientific satellite into space and the crew deployed it, this time for global change research. The Upper Atmosphere Research Satellite (UARS) was the first major element of NASA's Mission to Planet Earth, a global study that continued for more than twenty years.

In the mid- to late 1980s, scientists recommended that NASA begin to study Earth as thoroughly as it studied other planets. Amid growing concern about depletion of the ozone layer and the effects of human activity on the environment, a satellite to study the upper atmosphere won approval. This research soon evolved into a multi-satellite Earth observing system.

Ten different instruments to study the atmosphere as a complete environmental system were mounted on the UARS. Each one was tuned to investigate questions about the chemistry, physics, or dynamics of the upper atmosphere. What gases were there and in what concentrations? How did they mix and move around? How much energy did the sun add, and how constant or variable was that input? How did the middle and lower atmosphere respond to changes in the upper atmosphere, and vice versa?

The UARS was an important early step toward understanding global change; its measurements and data fed into predictions and informed decision making. The satellite far exceeded the planned three years in orbit; it operated for fourteen years until turned off in 2005 and finally deorbited in 2011. Its high-inclination orbit, suitable for studying global change, covered most of the Earth and gave it visibility to the Arctic and Antarctic regions.

During the mission, the crew also managed various experiments in the middeck, some making repeat appearances on the shuttle—protein

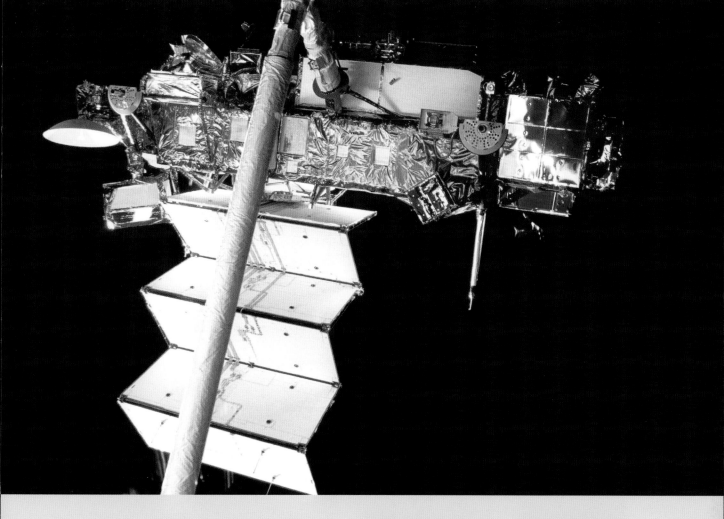

crystal growth, polymer formation, radiation monitoring, and others. Eight laboratory rodents came along for a study of the effects of weightlessness. The crew also tested, for the first time in space, a digital camera (a modified Nikon F4) to explore the advantages of electronic imaging over film.

In a rare occurrence, *Discovery* launched nine days *earlier* than projected and almost made the first night landing in Florida, but weather concerns there forced a return to California instead. Having already started Ulysses on its journey into interplanetary space and delivered the Hubble Space Telescope to observe the most distant reaches of the universe, *Discovery* this time started a series of missions to understand the environment of our home planet.

Above: Still held by the orbiter's long arm, the Upper Atmosphere Research Satellite unfolded its large solar panel before being released into orbit.

Right: In a high-inclination orbit, the shuttle traveled farther from the equator and crews could photograph more remote regions, such as the Antarctic ice shelf, seen here fraying along the edges into ribbons of sea ice.

STS-42: JANUARY 22–30, 1992

8 : 1 : 14
DAYS HRS MINS

129 | 57°

186 M
162 NM
(300 KM)

0 : 0
HRS MINS

0

EDWARDS
AIR FORCE BASE

45TH SPACE SHUTTLE MISSION

COMMANDER

- Ronald J. Grabe, USAF, his 3rd of 4 flights

PILOT

- Stephen S. Oswald, USN, his 1st of 3 flights, 2 on *Discovery*

MISSION SPECIALISTS

- Norman E. Thagard, medical doctor, his 4th of 5 Space Shuttle flights, plus a Soyuz flight and stay on Mir

- William F. Readdy, USN, aerospace engineer, his 1st of 3 flights, 2 on *Discovery*

- David C. Hilmers, USMC, electrical engineer, his last of 4 flights, 2 on *Discovery*

PAYLOAD SPECIALISTS

- Roberta L. Bondar, CSA, neurologist, her only flight

- Ulf D. Merbold, ESA, physicist, his 2nd shuttle flight, plus 2 Soyuz flights, and a stay on Mir

Star clusters denote STS-42, and the gold star pays tribute to crewmember Manley L. "Sonny" Carter Jr., who died in an airline crash several months before the mission. The orbiter is depicted flying nose-up, tail-to-Earth for microgravity research.

From left to right: Stephen S. Oswald, Roberta L. Bondar, Norman E. Thagard, Ronald J. Grabe, David C. Hilmers, Ulf D. Merbold, and William F. Readdy.

International Microgravity Laboratory

Discovery flew the first and last of eight shuttle missions in a year that marked the five hundredth anniversary of Columbus's voyages of exploration and discovery. By 1992, most of the scheduled satellite deployment and Department of Defense missions had been accomplished, and the shuttle missions of this decade increasingly were dedicated to scientific research. This time, *Discovery* flew with the Spacelab module and a two-shift crew working around the clock on microgravity science.

For this mission, *Discovery* stayed in a tail-to-Earth attitude that held its position with less need for corrective thruster firings. Microgravity research requires a very stable, smooth ride and a quiet environment without vibrations that can disturb or ruin delicate experiments. Vehicle maneuvers, crew activity, and even coughs and sneezes are jarring.

This first designated International Microgravity Lab mission had an international crew and experiments from about two hundred scientists in twenty countries. Neurologist Roberta Bondar became Canada's first woman in space, and German Ulf Merbold made his second flight; as a member of the Spacelab 1 crew in 1983, he had been the first European Space Agency astronaut and first non-American to fly on a U.S. spacecraft. Altogether, some fifty laboratory investigations tested the effects of microgravity on plants, animals, humans, and materials. In addition, ten Get Away Special canisters in the payload bay held automated experiments, including the first shuttle payload from China.

A Biorack with a properly controlled environment housed many of the life science experiments. Plant research focused on germination of seeds and growth of plant cells and seedlings, looking for any

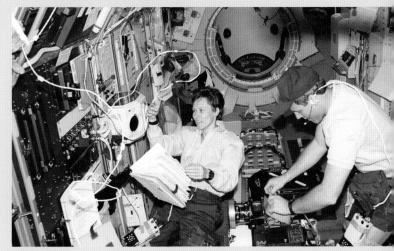

changes in growth rate or stem development. Oats, wheat, lentils, carrot cells, slime mold, and yeast spent a week in microgravity. Animal-based experiments included shrimp and fruit fly eggs, nematodes, bacteria, and cell cultures from frogs, mice, and humans. These experiments addressed radiation exposure, cell growth, genetics, and gravity-sensing mechanisms. Materials processing research included crystal growth experiments with organic and inorganic substances in a quest for larger, more uniform specimens.

Investigations of the human body's complex physiological responses, or adaptation, to microgravity tapped crew members to serve as test subjects. Several experiments involved a rotating chair and devices that measured a person's visual, sensation, and perception responses to different movement patterns. For the first time, the crew took the IMAX® camera into the Spacelab module to record research activities there, and some of their footage appeared in the IMAX® production *Destiny in Space* (1993).

After this mission, *Discovery* went out of service until late in the year for a major overhaul and upgrades. For its next mission, a modernized ship went to the launch pad.

Canadian Roberta L. Bondar works at the Biorack while pilot Stephen S. Oswald changes an IMAX® film magazine. Management of cords, manuals, and everything that is not anchored is a challenge in weightlessness.

Mission Specialist David C. Hilmers participates in a rotating chair experiment that spins him around or back and forth—on his side, back, or upright—to test his sensory perceptions of motion in weightlessness.

STS-53: DECEMBER 2–9, 1992

7 : 7 : 19
DAYS   HRS   MINS

116 | 57°

235 M
204 NM
(378 KM)

0 : 0
HRS   MINS

0

EDWARDS
AIR FORCE BASE

15TH *DISCOVERY* MISSION

COMMANDER

- David M. Walker, USN, his 3rd of 4 flights, 2 on *Discovery*

PILOT

- Robert D. Cabana, USMC, his 2nd of 4 flights, 2 on *Discovery*

MISSION SPECIALISTS

- Guion S. (Guy) Bluford Jr., USAF, aerospace engineer, his last of 4 flights, 2 on *Discovery*

- James S. Voss, USA, aerospace engineer, his 2nd of 6 shuttle flights, 3 on *Discovery*, and an ISS expedition

- Michael R. (Rich) Clifford, USA, aerospace engineer, his 1st of 3 flights

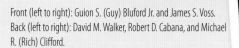

In a pentagon, the orbiter ascends on the symbol of the Astronaut Office, superimposed on five stars and three stripes for STS-53.

Front (left to right): Guion S. (Guy) Bluford Jr. and James S. Voss. Back (left to right): David M. Walker, Robert D. Cabana, and Michael R. (Rich) Clifford.

Last Department of Defense Shuttle Mission

Discovery's fifteenth flight, the last shuttle mission dedicated to Department of Defense purposes, closed the busiest shuttle year in space since nine missions flew in 1985. A classified primary payload—presumed to be a reconnaissance or defense communications satellite—two unclassified Get Away Special payloads in the bay, and nine middeck experiments occupied the all-military NASA crew on this weeklong mission.

The publicly disclosed parts of the mission included medical studies of the effects of microgravity on cells from bone tissue, muscles, and blood and measurements of radiation levels in the crew cabin. A variety of technology experiments flown on other missions were flown again to collect further data. These included a heat pipe cooling experiment, an attempted release of simulated space debris to test tracking systems on the ground (stalled by a dead battery), and a study of cloud formations from different angles. A laser detector made its first flight to test a tracking system on the ground, and the crew tried a simulated refueling experiment by transferring fluids in weightlessness. Some of these technology experiments were designed with space station engineering in mind.

As this was the last of ten dedicated Department of Defense missions, it was fitting, although perhaps coincidental, that the crew represented the navy, marine corps, air force, and army. From 1980 through 1985, the military services selected and trained a cadre of manned spaceflight engineers to serve as payload specialists on classified missions. Of the thirty-three selected, only two were assigned to flights—air force officers Gary E. Payton on *Discovery* STS-51C and William A. Pailes on *Atlantis* STS-51J, the first two DOD missions. Thereafter, NASA astronauts with military credentials filled all crew positions on DOD missions.

At mission's end, both the head of NASA and the assistant secretary of the air force for space remarked on the long and productive interagency cooperation realized in the Department of Defense missions. There were no more classified shuttle flights.

Above: This view of *Discovery*'s December 1992 launch is from the top of the Vehicle Assembly Building at Kennedy Space Center, once the tallest building in Florida.

Left: The STS-53 landing marked *Discovery*'s first use of a drag chute, installed during its 1992 modification.

9 : 6 : 8
DAYS HRS MINS

184 M
160 NM
(296 KM)

148 | 57°

0 : 0
HRS MINS

0

KENNEDY SPACE CENTER

COMMANDER

- Kenneth D. Cameron, USMC, his 2nd of 3 flights

PILOT

- Stephen S. Oswald, USN, his 2nd of 3 flights, 2 on *Discovery*

MISSION SPECIALISTS

- C. Michael Foale, astrophysicist, his 2nd of 6 shuttle flights, 3 on *Discovery*, and stays on Mir and the ISS

- Kenneth D. Cockrell, USN, aeronautical systems engineer, his 1st of 5 flights

- Ellen Ochoa, electrical engineer, her 1st of 4 flights, 2 on *Discovery*, RMS operator

This Mission to Planet Earth is symbolized by the ATLAS 2 instruments mounted in the payload bay to study the sun and Earth's atmosphere.

From left to right: Kenneth D. Cockrell, Stephen S. Oswald, C. Michael Foale, Kenneth D. Cameron, and Ellen Ochoa.

ATLAS and SPARTAN Observatories

This Spacelab mission was kindred to several others in a coordinated study of the space environment around Earth. In 1991, the STS-48 *Discovery* crew deployed the Upper Atmosphere Research Satellite, the first in a series of flights called Mission to Planet Earth. The next year, *Atlantis* carried a platform bearing a suite of instruments collectively called ATLAS on the STS-45 mission. *Discovery* now brought the Atmospheric Laboratory for Applications in Space (ATLAS) up for a second round of observations, and the ATLAS payload would fly yet again on *Atlantis* in 1994 on the STS-66 mission.

ATLAS comprised seven remote sensing instruments that simultaneously measured in different ways solar energy output, atmospheric chemistry, and the presence of ozone. Amid concerns about depletion of the protective atmospheric ozone layer, scientists sought to understand better the relationships among these factors and to determine how changes in one affected the others. The high-inclination orbit of the three ATLAS missions covered most of the Earth, yielding data for comparison at different locales and seasons.

The ATLAS instruments were mounted in the payload bay for unobstructed observations, all but one sharing space on a Spacelab pallet. The crew managed the instruments' operations from the cabin and maneuvered the orbiter to point the cluster directly toward the sun or atmosphere at required angles or times. Sunrises and sunsets made prime opportunities for analyzing chemical elements in the illuminated layers of the atmosphere along the Earth's limb.

Another SPARTAN scientific satellite, this time outfitted to measure the solar wind velocity and observe the sun's corona, was on board. Lifted out of the orbiter by the manipulator arm, SPARTAN-201 flew solo for two days, free to maneuver on its own to carry out its program, before

retrieval for return to Earth. Mission Specialist Ellen Ochoa's duties included operating the robotic arm to handle SPARTAN; she made history on this flight as the first Hispanic woman in space.

The payload also included eight middeck experiments, one of which was a facility hosting more than thirty investigations of materials, cells, and tissue samples exposed to microgravity. The crew worked in two shifts around the clock to keep all experiments inside and all instruments outside running continuously. Staying in touch with researchers on the ground throughout the mission helped to ensure best results.

The ATLAS-2 mission and others in the Mission to Planet Earth series enabled researchers to put complementary instruments into the nearby space environment in an effort to decipher some of the natural influences and interactions there.

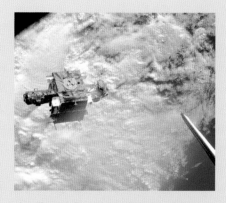

Below: Most of the ATLAS observatory is visible in the foreground of this payload bay view. SPARTAN is not visible because it is flying on its own, but it fits between ATLAS and the aft bulkhead where the flag is seen.

Above: Flying solo for two days, SPARTAN-201 carried out its independent observing program focused on the sun and solar wind.

9 : 20 : 11
DAYS **HRS** **MINS**

200 M
174 NM
(322 KM)

157 | 28.5°

7 : 5
HRS **MINS**

1

KENNEDY SPACE
CENTER

COMMANDER

- Frank L. Culbertson Jr., USN, his 2nd of 4 shuttle flights, 2 on *Discovery*, plus an ISS expedition

PILOT

- William F. Readdy, USN, his 2nd of 3 flights, 2 on *Discovery*

MISSION SPECIALISTS

- James H. Newman, physicist, his 1st of 4 flights, 1 EVA

- Daniel W. Bursch, USN, engineer, his 1st of 5 shuttle flights, plus an ISS expedition, RMS operator

- Carl E. Walz, USAF, physicist, his 1st of 5 shuttle flights, plus an ISS expedition, 1 EVA

Five large stars for the crewmembers and a single gold star symbolize the numerals for STS-51. The gold star and its rays also represent the ACTS satellite. The shape with colors of the German flag represents the SPAS satellite, and the constellation Orion refers to the ORFEUS-SPAS stellar observing program.

Satellites, Science, and Spacewalking

Discovery's STS-51 mission seemed like three missions in one: satellite deployment, scientific research, and extravehicular activity. Once again, this mission showcased the versatility of the shuttle vehicle and crew.

But before that could happen, *Discovery* had to get off the ground. Since the crew first boarded the vehicle in July, their launch was called off three times for technical reasons after they were strapped in, once at the T-3 seconds mark after the main engines had ignited. This was the second on-pad abort for *Discovery* (see STS-41D), a nerve-wracking halt less than a breath before liftoff. In addition, NASA postponed the launch date five times for other reasons, once to reduce the risk of meteoroid damage during the peak of the Perseid meteor shower in August.

Finally in orbit, the crew on the first day deployed the Advanced Communications Technology Satellite (ACTS), attached to a TOS solid rocket upper stage—the only time it was used with a shuttle payload. The ACTS was an experimental test bed for new switching, processing, and antenna technologies that would speed up and increase capacity for voice and data transmissions. The satellite's spring-loaded release from the payload bay made a bigger "bang" than expected when its explosive hold-down cords detonated, causing minor damage to thermal blankets on the aft bulkhead of the bay.

Dan Bursch used the manipulator arm for the next day's deployment of the ORFEUS-SPAS retrievable satellite for the first in a series of joint German–United States ASTRO-SPAS astronomy missions. ORFEUS consisted of a 1-meter telescope and detectors tuned to far and

From left to right: Frank L. Culbertson Jr., Daniel W. Bursch, Carl E. Walz, William F. Readdy, and James H. Newman.

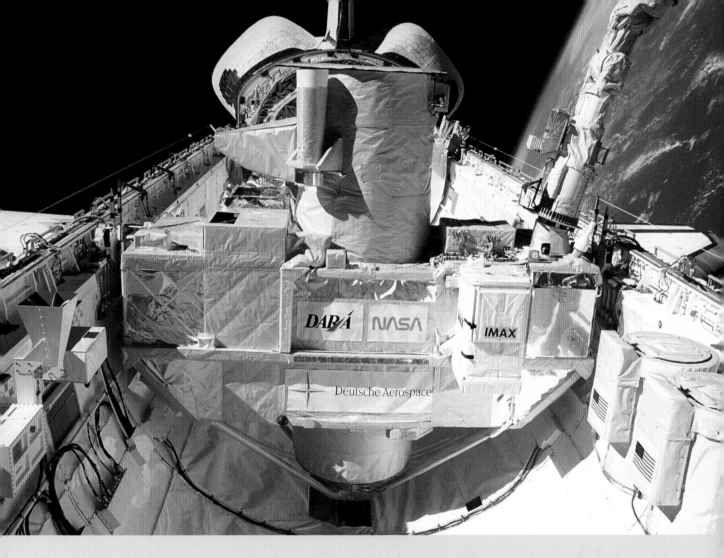

extreme ultraviolet radiation, signatures of the birth and death of stars. This was the fourth of seven flights for the SPAS free-flyer and its second time on a *Discovery* mission (see STS-39). Bursch, Frank Culbertson, and Bill Readdy choreographed the rendezvous and retrieval maneuvers six days later.

Meanwhile, the crew carried out seven middeck experiments, including one focused on chromosome changes in plant cells and some biomedical tests, and they tested a GPS receiver for the first time in space. While they used the onboard IMAX® camera to record mission activities, a remotely operated IMAX® camera on the SPAS captured images of the Earth and *Discovery*. Footage from the cameras appeared in two IMAX® feature films, *Destiny in Space* and *Space Station 3D*.

On the fifth day, Carl Walz and Jim Newman went on a test spacewalk to evaluate tools and procedures to be used in the upcoming first Hubble Space Telescope servicing mission. Besides checking out wrenches, tethers, and a portable foot restraint, they evaluated how much force it took to do certain operations in space compared to their underwater training and how much mobility their restraint devices allowed. They did not use the remote manipulator system, instead keeping it ready for the next day's SPAS retrieval.

The STS-51 timeline, as for all shuttle missions, left little time for rest. Crews always strove to accomplish as much as possible in their limited time in space. The more diverse the mission tasks, the greater the challenge to complete the timeline. This three-in-one mission met those expectations.

Held by the remote manipulator arm, the SPAS satellite with the ORFEUS telescope suite and IMAX® camera is returned to the payload bay.

223 M
194 NM
(359 KM)

130 | 57°

KENNEDY SPACE
CENTER

0

60TH SPACE SHUTTLE MISSION

COMMANDER

- Charles F. Bolden Jr., USMC, his last of 4 flights, 2 on *Discovery*

PILOT

- Kenneth S. Reightler Jr., USN, his last of 2 flights, both on *Discovery*

MISSION SPECIALISTS

- N. Jan Davis, mechanical engineer, her 2nd of 3 flights, 2 on *Discovery*, RMS operator

- Ronald M. Sega, USAF, electrical engineer, his 1st of 2 flights

- Franklin R. Chang-Diaz, physicist, his 4th of 7 flights, 2 on *Discovery*

- Sergei K. Krikalev, RSA, mechanical engineer, his 1st of 3 shuttle flights, plus 2 stays on Mir and 2 ISS expeditions, RMS operator

Wings depicted as United States and Russian flags symbolize partnership in spaceflight for the first time since the 1975 Apollo-Soyuz mission. The primary payload, the Wake Shield Facility, is featured on the orbiter's arm.

Clockwise from front left: Kenneth S. Reightler Jr., Franklin R. Chang-Diaz, Ronald M. Sega, Sergei K. Kirkalev, N. Jan Davis, and Charles F. Bolden Jr.

First Cosmonaut on the Shuttle

Shuttle missions often did not fly in their assigned numerical order, but STS-60 was indeed the sixtieth shuttle mission and the first in a seven-mission year. This was a transitional mission, the first in a series of eleven cooperative flights involving the United States and Russia and the second in a series of flights of the SPACEHAB commercial research module. This mission also featured a large technology experiment—the Wake Shield Facility—held overboard in orbit on the long remote manipulator arm.

In 1992, President George H. W. Bush and President Boris Yeltsin signed an agreement on cooperation in space between the United States and the Russian Federation. At a minimum, a cosmonaut would fly on the Space Shuttle and an astronaut would spend time on the Russian space station Mir, and the shuttle would dock with Mir once. The agreement soon expanded to authorize up to ten more cooperative Shuttle-Mir missions to build experience for joint operation of the future International Space Station. STS-60 inaugurated the program, and Sergei Krikalev, veteran of two Mir missions, became the first Russian cosmonaut to launch on an American spacecraft and serve on its crew.

As a mission specialist, Krikalev actively engaged in science activities on the flight and also served with payload commander Jan Davis as a robotic arm operator. Davis had primary responsibility for maneuvering the Wake Shield Facility, a 12-foot diameter (3.7-meter) curved, stainless-steel disk with attached experiments. When held beside the shuttle on the extended remote manipulator arm or released as a free-flyer, the shield pushed stray atoms aside and created an ultra-vacuum in its own wake—possibly making the back side of the disk an ideal place for vacuum-sensitive production processes. Technical problems kept the Wake Shield Facility on the arm this time, but it was released twice on later *Endeavour* and *Columbia* missions. The flights succeeded in producing samples of very high-quality crystalline semiconductor thin films.

SPACEHAB, a habitable commercial module similar to the Spacelab module produced for NASA by the European Space Agency, fit into the payload bay just behind the airlock hatch, enabling the crew to pass easily back and forth from the middeck. SPACEHAB expanded the space available for experiments and supplies on eight Shuttle-Mir missions and five International Space Station missions. This time, SPACEHAB carried various crew-tended life science and materials science experiments, small centrifuges, a freezer for blood and urine samples, a compact furnace for processing materials, and crystal growth devices. Other experiments occupied GAS canisters in the payload bay.

A year after STS-60, the first Shuttle-Mir rendezvous occurred—also a *Discovery* mission—followed by nine successful docking missions. Other cosmonauts became members of shuttle crews, and by 1998 seven astronauts had lived on Mir. The modern era of American and Russian cooperation in space began on this *Discovery* mission.

Above: As a member of the STS-60 crew, cosmonaut Sergei K. Krikalev was the first Russian to fly on a U.S. spacecraft and on the first in a series of joint U.S.-Russia missions from 1994 through 1998. Here he is on the aft flight deck talking with students via amateur radio equipment while preparing to use a camcorder.

Left: The front or ram side of the shield facility generated an ultrapure-vacuum in its wake as the large disk moved through space like a plate on edge. The orbiter's arm grasped a fixture on the front side to position the shield.

10:22:49
DAYS HRS MINS

176 | 57°

162 M
141 NM
(261 KM)

6:51
HRS MINS

1

EDWARDS
AIR FORCE BASE

COMMANDER

- Richard N. Richards, USN, his last of 4 flights, 2 on *Discovery*

PILOT

- L. Blaine Hammond Jr., USAF, his last of 2 flights, both on *Discovery*

MISSION SPECIALISTS

- Jerry M. Linenger, USN, medical scientist, his 1st of 3 shuttle flights, and a stay on Mir

- Susan J. Helms, USAF, engineer, her 1st of 6 shuttle flights, 3 on *Discovery*, and an ISS expedition, RMS operator

- Carl J. Meade, USAF, electronics engineer, his last of 3 flights, 1 EVA

- Mark C. Lee, USAF, mechanical engineer, his 3rd of 4 flights, 2 on *Discovery*; 1 EVA

The astronaut corps symbol—a star with three rays passing through a circle—also represents the three-wavelength laser experiment. Two EVA astronauts holding the symbol are depicted with their new SAFER jetpacks, and the orbiter's arm holds the SPARTAN satellite. Gold (USN) and silver (USAF) stars identify the service of each crewmember.

Trying Something New

Discovery's second mission in 1994 accomplished several firsts by testing new technologies. Simultaneously, some familiar payloads made repeat flights. STS-64 mixed novelty for the future and continuity with the past.

Flying another Mission to Planet Earth, *Discovery* had a new primary payload: a LIDAR (Light Detection and Ranging) instrument, the first laser system used in space for atmospheric research. LIDAR is a type of "radar" that beams laser light pulses instead of radio waves; their reflections reveal the structure of clouds and storm systems as well as the presence of dust, smoke, and pollutants in the atmosphere.

The crew operated the large device mounted in the payload bay from the aft flight deck control station and reoriented the orbiter as necessary to test the technology at various angles to the atmosphere. This LITE (LIDAR In-Space Technology Experiment) test program ran very successfully and produced atmospheric data over various parts of the globe for comparison with data collected from the ground and aircraft in a multinational research effort.

The EVA crew, Mark Lee and Carl Meade, achieved another milestone: the first untethered spacewalk in ten years (and the last one in the shuttle program). They tested a new propulsion backpack, the SAFER, meant for self-rescue in case of an emergency. Much smaller than the Manned Maneuvering Unit last used in 1984, the SAFER fit below the suit's life support systems backpack like a fanny pack. Using a joystick controller, an astronaut could return to safety if ever unmoored from the shuttle or space station.

EVA astronauts wore the SAFER through the rest of the shuttle era and on International Space Station EVAs, but as of 2014 no emergency

Front (left to right): L. Blaine Hammond Jr., Richard N. Richards, and Susan J. Helms. Back (left to right): Mark C. Lee, Jerry M. Linenger, and Carl J. Meade.

Left: No longer connected to the orbiter by a safety tether, Mark C. Lee tests the new SAFER self-rescue device (the lower box on his back) by operating the controller with his right hand.

Below: Flying bay down toward Earth and at a lower altitude than usual, *Discovery*'s crew operated the LITE laser technology experiment mounted in the payload bay to study the atmosphere.

had occurred to prompt its use. Lee and Meade also tested some EVA tools and an electronic checklist alternative to the paper checklist worn on the EVA suit's sleeve.

Two STS-64 crewmembers wore for the first time the fully pressurized ACES suits that gradually replaced the partial pressure Launch-Entry Suits (LES) first worn on *Discovery*'s STS-26 mission. Orange like its predecessor "pumpkin suit," the ACES had gloves and helmet with locking metal rings, making it a completely sealed protective suit for the most dangerous phases of flight.

In the payload bay, a device called ROMPS made its debut as the first U.S. robotic system flown on the shuttle; it was a mechanism for processing semiconductor materials. Another new experiment called SPIFEX held a sensor on a long extension attached to the orbiter's arm to study the effects of exhaust from the vehicle's thruster firings, a topic of interest for operations near the Hubble Space Telescope or a space station. The SPARTAN 201 free-flyer on its second mission was deployed for two days to observe the solar wind and the sun's corona. In the middeck, repeat experiments included a study of how flames spread in microgravity (seventh flight) and a plant biology study (second flight).

The STS-64 crew enjoyed two bonus days in space, one to complete more science and the other a weather-related delay. With its mix of familiar and new technologies, this mission tested remote sensing techniques and helped pave the way to the space station.

STS-63: FEBRUARY 3–11, 1995

8 : 6 : 28
DAYS HRS MINS

4 : 38
HRS MINS

1

246 M
214 NM
(396 KM)

129 | 51.6°

KENNEDY SPACE CENTER

20TH *DISCOVERY* MISSION

COMMANDER
- James D. Wetherbee, USN, his 3rd of 6 flights, 2 on *Discovery*

PILOT
- Eileen M. Collins, USAF, her 1st of 4 flights, 2 on *Discovery*

MISSION SPECIALISTS
- Bernard A. Harris Jr., medical doctor, his last of 2 flights, 1 EVA
- C. Michael Foale, astrophysicist, his 3rd of 6 shuttle flights, 3 on *Discovery*, plus 2 Soyuz flights, a stay on Mir, and an ISS expedition, 1 EVA
- Janice E. Voss, engineer, her 2nd of 5 flights, RMS operator
- Vladimir G. Titov, RSA, his 1st of 2 shuttle flights; also 3 Soyuz flights and a stay on Mir, RMS operator

Six rays of the sun and three stars stand for STS-63, the first rendezvous of the U.S. Space Shuttle with the Russian space station Mir. The flat-topped SPACEHAB and triangular SPARTAN are represented in the open payload bay.

Front (left to right): Janice E. Voss. Eileen M. Collins, James D. Wetherbee, and Vladimir G. Titov. Back (left to right): Bernard A. Harris Jr., and C. Michael Foale.

Multiple Firsts

The first 1995 shuttle launch marked *Discovery*'s twentieth trip into orbit and the first U.S. mission to the Russian space station Mir. That and another historic event made headlines: Eileen Collins became the first woman to pilot a U.S. spacecraft. Yet another first occurred near the end of the mission, when Bernard Harris became the first African American to do an EVA, a spacewalk. And for the second time, a shuttle crew included a cosmonaut, now Vladimir Titov, who had already spent a year on Mir. This mission was destined for the history books.

To match Mir's orbit, the shuttle's trajectory for the first time was inclined 51.6° to the equator, and the launch window was a tight five minutes. On the third day in space, Jim Wetherbee and Collins began the thruster firings that would bring *Discovery* into Mir's neighborhood. However, three of the forty-four thrusters for fine maneuvering in space were not functioning properly, which put the rendezvous in jeopardy. After analysis yielded a different thruster strategy, the Russian and American mission managers agreed to the close approach.

Wetherbee then flew *Discovery* within 37 feet (11 meters) of Mir, while Titov talked by ship-to-ship radio with the Mir crew. After station-keeping there for fifteen minutes, *Discovery* backed away to 400 feet (122 meters) and for the next hour flew a loop and a half around the space station. These maneuvers demonstrated techniques to be used in the upcoming series of nine Shuttle-Mir docking missions.

Although the Mir rendezvous was the main objective, STS-63 carried two major payloads—SPACEHAB on its third flight and SPARTAN 204 on its first flight. The SPACEHAB laboratory module held twenty experiments in biotechnology and materials science. SPARTAN 204 served two purposes. Operating at the end of the orbiter's arm, it observed thruster firings and the glow that appeared along the surface

of the vehicle. Released as a free-flyer for two days, SPARTAN observed far ultraviolet radiation from the interstellar medium, the seemingly empty space between stars.

Familiar repeat payloads included the IMAX® camera, a combustion experiment, an astroculture plant growth facility, and an orbital debris tracking experiment. The crew also experimented with a new robotic device called Charlotte to perform simple tasks such as operating switches and moving experiment samples.

The spacewalk occurred on the seventh day, when Mike Foale and Bernard Harris went outside to test EVA suit refinements meant to keep astronauts warmer when working in darkness or beyond the radiated heat of the payload bay. They also intended to do a mass handling exercise with the 2,500-pound (1,134-kg) SPARTAN, but they reported that they were growing very cold. That task was curtailed, but the EVA was notable as the first spacewalk by an African American.

Eileen Collins went on to pilot an *Atlantis* mission to Mir in 1997 and command a 1999 *Columbia* mission and *Discovery*'s return-to-flight-mission in 2005. She was the first American woman selected to be a pilot astronaut and the first woman to pilot and command shuttle missions. She became eligible to be a space pilot as soon as it was possible for a woman to do so, when air force aviation command assignments and test pilot school opened to women. After her, only two other women pilots served on the shuttle, one of whom also commanded *Discovery*.

Because *Discovery* did not dock with the Russian space station, some called STS-63 the "near Mir" mission. However, both spacecraft commanders well recognized the symbolic significance of this close encounter. During the rendezvous, Wetherbee remarked that "we are bringing our nations closer together" and "we will lead our world into the next millennium," to which Alexander Viktorenko responded, "We are one." *Discovery* would not visit Mir again until 1998, when it made history again on the final docking mission.

Cosmonaut Vladimir G. Titov works with a biotechnology experiment inside the SPACEHAB module.

Cosmonaut Valery Polyakov watches *Discovery* from a window on Mir during the rendezvous.

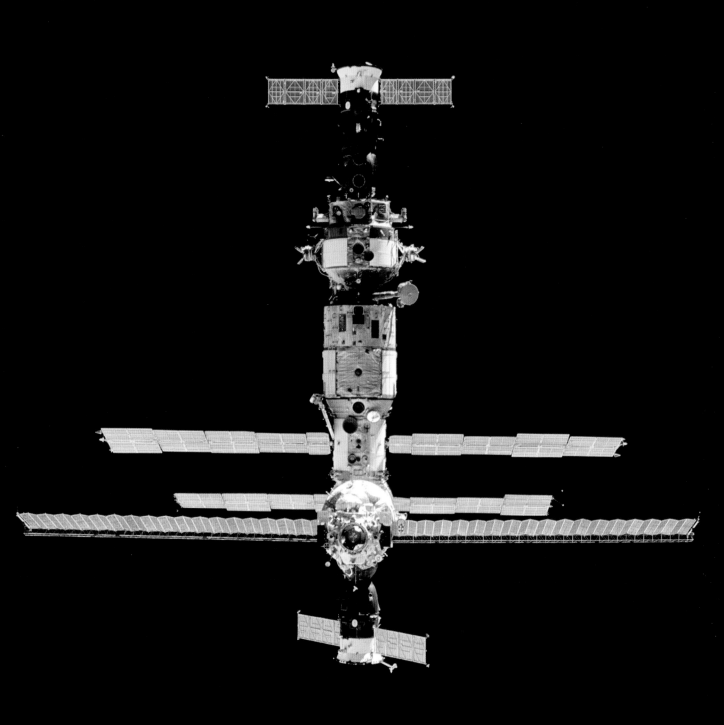

Above: Bernard A. Harris Jr. and C. Michael Foale (red stripes) move around on the remote manipulator arm operated by Janice E. Voss.

Left: First woman in the pilot's seat: Eileen Collins makes history.

Opposite: The close encounter gave *Discovery*'s crew opportunities to photograph the Mir space station, often said to resemble a dragonfly, in detail from various angles and distances.

8 : 22 : 20
DAYS   HRS   MINS

191 M
166 NM
(307 KM)

143 | 28.5°

0 : 0
HRS   MINS

0

KENNEDY SPACE
CENTER

70TH SPACE SHUTTLE MISSION

COMMANDER

- Terence T. (Tom) Henricks, USAF, his 3rd of 4 flights

PILOT

- Kevin R. Kregel, USAF, his 1st of 4 flights

MISSION SPECIALISTS

- Donald A. Thomas, materials scientist, his 2nd of 4 flights

- Nancy J. Currie, USA, industrial engineer, her 2nd of 4 flights

- Mary Ellen Weber, physical chemist, her 1st of 2 flights

Shaped like the block letter *O* symbol for Ohio State University to recognize the all-Ohio crew, this design depicts as stars the three TDRS satellites deployed from *Discovery*.

End of *Discovery's* First Term

Thanks to a woodpecker's determined effort to make a nest on the vehicle, this *Discovery* mission was the only shuttle flight delayed by wildlife. It also was the only mission with an all-Ohio crew. Its real importance, though, was deployment of the sixth NASA Tracking and Data Relay Satellite and the last mission of *Discovery* in its early configuration.

Less than two weeks before the planned launch date, NASA discovered damage to the foam insulation covering the external tank. Closer inspection revealed about two hundred holes ranging in size from tiny to about 4 inches (10 centimeters) across. Apparently, a single male Northern Flicker woodpecker drilled these holes in a futile effort to create an attractive nest for a female. Repair at the launch pad was impractical, so the *Discovery* stack was moved back to the Vehicle Assembly Building. As a result, the STS-71 first docking mission to Mir flew ahead of STS-70 as the one hundredth mission in U.S. human spaceflight history. *Discovery* launched six days after *Atlantis* returned, the briefest-ever interval between missions.

Coincidentally, four of the five astronauts assigned to this crew were born in Ohio, and at the commander's suggestion the governor declared pilot Kevin Kregel an honorary citizen of the state. The crew parlayed the "all-Ohio mission" into recognition of Ohio's many contributions to aerospace and astronautics from the Wright brothers to more than twenty Ohio astronauts since John Glenn became the first.

The STS-70 crew achieved their primary objective six hours after launch by deploying the sixth satellite to complete the initial TDRS constellation. This was the last TDRS taken to space by the shuttle; Atlas

From left to right: Kevin R. Kregel, Nancy J. Currie, Terence T. (Tom) Henricks, Mary Ellen Weber, and Donald A. Thomas.

Above: An unexpected rollback for repairs and a second trip to the pad delayed the STS-70 launch by five weeks, causing *Discovery* to miss making the one hundredth U.S. human spaceflight mission.

Below: Light spots on the external tank are patched woodpecker holes. Orange predator-eye scare balloons hang nearby to discourage the bird's return.

rockets launched later versions still used for continuous communication with spacecraft.

The crew then spent the next eight days in orbit doing research, again using the shuttle middeck as a laboratory for life science and materials processing experiments in microgravity. A bioreactor used to grow cancer cells into tissue cultures for study performed well, as did investigations of plant cell division, animal anatomy and physiology, and protein crystal growth. The crew also tested their visual acuity in response to reports of eyesight changes in space.

On the aft flight deck, the crew worked with a new device called HERCULES, a video camera coupled with an inertial navigation system so it could record latitude and longitude data on images. Aligning it properly by star sightings proved difficult, but the astronauts worked in concert with the camera team on the ground until they succeeded. A set of observations through the windows— nicknamed WINDEX—monitored atmospheric airglow, shuttle glow, and the effects of thruster firings.

NASA inaugurated its new flight control room in the Mission Control Center for most of the STS-70 orbital operations, and a Space Shuttle main engine with an improved design flew for the first time on this *Discovery* mission. As the shuttle program kept moving forward, *Discovery* took a nine-month break for modifications at the assembly plant in Palmdale, California. Having completed twenty-one missions in its first ten years of flight (none in 1986–87), *Discovery* would return to flight in 1997 with a new airlock and a docking adapter, ready for future missions to a space station.

STS-82

9 : 23 : 37
DAYS HRS MINS

386 M
335 NM
(620 KM)

150 | 28.5°

33 : 11
HRS MINS

5

KENNEDY SPACE
CENTER

COMMANDER

- Kenneth D. Bowersox, USN, his 4th of 5 shuttle flights and an ISS expedition

PILOT

- Scott J. Horowitz, USAF, his 2nd of 4 flights, 2 on *Discovery*

MISSION SPECIALISTS

- Joseph R. Tanner, USN, mechanical engineer, his 2nd of 4 flights, 2 EVAs

- Steven A. Hawley, astronomer/astrophysicist, his 4th of 5 flights, 3 on *Discovery*, RMS operator

- Gregory J. Harbaugh, engineer, his last of 4 flights, 2 on *Discovery*, 2 EVAs

- Mark C. Lee, USAF, mechanical engineer, his last of 4 flights, 2 on *Discovery*, 3 EVAs

- Steven L. Smith, electrical engineer, his 2nd of 4 flights, 2 on *Discovery*, 3 EVAs

The design depicts the Hubble Space Telescope pointed toward deep space as viewed from the approaching shuttle. EVA crew names are grouped above the spiral galaxy.

Hubble Space Telescope Servicing Missions

Discovery delivered the Hubble Space Telescope in 1990 and returned twice on the second and third of five servicing missions. These two shuttle crews collectively installed two new scientific instruments and replaced or added twenty telescope systems components to update the orbital observatory and restore it to mint working condition.

The goals for the 1997 mission were to exchange two scientific instruments and replace ten components that had degraded from seven years of operating in space. Within three days of launch, the crew reached the telescope and Steve Hawley, who had originally deployed it, used the orbiter's arm to grasp Hubble and gingerly bring it aboard for servicing. Five days of intense work outside by two spacewalker teams followed. Mark Lee and Steve Smith alternated EVA days with Greg Harbaugh and Joe Tanner, and Hawley operated the arm for both teams. Designed to be serviced in space, most equipment was readily accessible in bays behind the large and small doors that circled the telescope.

On STS-82, Lee and Smith first removed two telephone-booth-size spectrographs and installed two new scientific instruments that extended Hubble's ability to study the universe across a broader

Left: Front (left to right): Kenneth D. Bowersox, Steven A. Hawley, and Scott J. Horowitz. Back (left to right): Joseph R. Tanner, Gregory J. Harbaugh, Mark C. Lee, and Steven L. Smith.

Opposite: Aided by the orbiter's arm, Steven L. Smith moves one of the large scientific instruments between its place inside the telescope and a storage fixture in the payload bay.

STS-103

 **7 : 23 : 10**
DAYS HRS MINS

 119 | 28.5°

 380 M
330 NM
(611 KM)

24 : 25
HRS MINS

3

KENNEDY SPACE CENTER

COMMANDER

- Curtis L. Brown Jr., USAF, his last of 6 flights, 3 on *Discovery*

PILOT

- Scott J. Kelly, USN, his first of 2 shuttle flights and an ISS expedition

MISSION SPECIALISTS

- Steven L. Smith, electrical engineer, his 3rd of 4 flights, 2 on *Discovery*, 2 EVAs

- Jean-François A. Clervoy, ESA, engineer, his last of 3 flights, RMS operator

- John M. Grunsfeld, physicist, his 3rd of 5 flights, 2 EVAs

- C. Michael Foale, astrophysicist, his last of 6 shuttle flights, 3 on *Discovery*, a stay on Mir, an ISS expedition, 1 EVA

- Claude Nicollier, ESA, astrophysicist, his last of 4 space flights, 1 EVA

Its solar arrays illuminated by the sun, the telescope rests on two intersecting lines that represent precise attitude control, to be restored by this servicing mission.

From left to right: C. Michael Foale, Claude Nicollier, Scott J. Kelly, Curtis L. Brown Jr., Jean-François A. Clervoy, John M. Grunsfeld, and Steven L. Smith.

European Space Agency astronaut Claude Nicollier uses a power tool, one of more than 150 tools available for these servicing missions.

wavelength range and to make infrared images. The second and third EVAs dealt with practical matters as Tanner and Harbaugh, then Lee and Smith, replaced components in the telescope's pointing system and a data recorder. The last two EVA days focused on replacing solar array electronics and adding another layer of thermal insulation on the exterior where the silver-color blankets had frayed from long exposure to the harsh space environment.

The 1999 servicing mission occurred two years earlier than planned, prompted by the failure of three of the six rate sensor gyroscope units needed for telescope pointing and attitude control. Hubble went into an idle "safe mode" and would not be used for observations until it was repaired. The STS-103 crew replaced the gyro units, another fine guidance sensor and data tape recorder, and a data transmitter. They also installed an advanced master computer and other new upgrades. Steve Smith and John Grunsfeld worked the first and third EVA days, and Michael Foale and Claude Nicollier worked the second, all lasting more than eight hours, among the longest EVAs in shuttle history. Jean-François Clervoy operated the orbiter's arm for the telescope retrieval, EVAs, and release into space.

As work progressed on the STS-82 mission, commander Ken Bowersox and pilot Scott Harbaugh flew *Discovery* to successively higher altitudes, giving Hubble a final boost to the record-high shuttle orbit. This pushed the record altitude set by *Discovery* on the STS-31 Hubble deployment mission into second place. STS-103 commander Curt Brown and pilot Scott Kelly attained the third highest shuttle orbit without a telescope reboost. No other orbiter surpassed these *Discovery* records.

Two more servicing missions followed in 2002 and 2009 to double the operational life of the Hubble Space Telescope. The Hubble servicing missions rank among the most complicated and most significant of all shuttle missions for that reason and another: They expanded EVA astronauts' ability to do intricate tasks and showed the value of human mental and physical dexterity in space. In servicing the telescope, astronauts proved that it was feasible to assemble and service a vastly larger and more complex structure: a space station.

Every *Discovery* mission after STS-103 went to the International Space Station.

Above: Secure on a foot restraint attached to the orbiter's arm, Gregory J. Harbaugh handles a fine guidance sensor the size of a baby grand piano. The base of the telescope looms behind him.

Left: Steven L. Smith and John M. Grunsfeld appear small beside the four-story telescope. They have opened the lower bay doors to replace failed gyroscope units.

11 : 20 : 28
DAYS HRS MINS

184 M
160 NM
(296 KM)

189 | 57°

0 : 0
HRS MINS

0

KENNEDY SPACE
CENTER

COMMANDER

• Curtis L. Brown Jr., USAF, his 4th of 6 flights, 3 on *Discovery*

PILOT

• Kent V. Rominger, USN, his 3rd of 5 flights, 2 on *Discovery*

MISSION SPECIALISTS

• N. Jan Davis, mechanical engineer,
her last of 3 flights, 2 on *Discovery*, RMS operator

• Robert L. Curbeam Jr., USN, aeronautics-astronautics
engineer, his 1st of 3 flights, 2 on *Discovery*

• Stephen K. Robinson, mechanical engineer,
his 1st of 4 flights, 3 on *Discovery*

PAYLOAD SPECIALIST

• Bjarni V. Tryggvason, CSA, engineer, his only flight

The design is filled with
symbols of the payloads and
subjects of research on this
mission's agenda.

Front (left to right): Kent V. Rominger and Curtis L. Brown Jr. Back (left to right):
Robert L. Curbeam Jr., Stephen K. Robinson, N. Jan Davis, and Bjarni V. Tryggvason.

Another Mission
to Planet Earth

Between the 1997 and 1999 Hubble Space Telescope servicing missions, *Discovery* flew four times on quite different missions. The first of these marked *Discovery*'s longest flight to date—almost twelve days—and another Mission to Planet Earth with a SPAS scientific satellite equipped to study the atmosphere.

The acronym CRISTA referred to a suite of infrared sensors on their second trip to space. Like other SPAS payloads, this was a cooperative venture of the German Space Agency and NASA. As on similar atmospheric research missions, *Discovery* flew in a high-inclination orbit for more global coverage. The satellite spent eight days as a free-flyer collecting data on the location and dynamics of ozone and other gases in concert with instrumented balloons, sounding rockets, and devices on the shuttle.

The crew also used the orbiter's arm to maneuver the SPAS to test a procedure planned for the first International Space Station assembly flight. In an EVA exercise, they practiced handling a small manipulator arm to be installed on Japan's portion of the station. Trials of space station assembly techniques had begun on shuttle missions in the 1980s, but now such practice had more urgency; the first space station modules would be launched in about a year. Flight plans for shuttle missions in the 1990s often included rehearsal for space station assembly tasks.

Discovery carried various experiments in the payload bay, some mounted on two Hitchhiker carriers. These were reusable support structures that provided power, control, and data transmission services for small payloads. To complement the CRISTA investigations, one

Hitchhiker held instruments tuned to record extreme ultraviolet radiation from the sun and celestial sources. The other held a mix of solar and avionics instruments, the latter being tested under NASA's "faster, better, cheaper" rubric for streamlining spaceflight.

Meanwhile, several middeck experiments flew again, including the bioreactor for culturing cancer cells and the protein crystal growth facility. Re-flying experiments enabled scientists and crew to refine research equipment and techniques in readiness for longer-term investigations on the space station. As always, the crew had enough cameras and photography sessions to capture visually and scientifically remarkable images of the Earth.

A shuttle mission focused on scientific research still served as a good opportunity for engineering evaluation of space hardware and procedures inside and outside the vehicle. *Discovery*'s last missions before 2000 helped prepare the way to the International Space Station.

Shooting photographs through the windows, the crew monitored super-typhoon Winnie, then covering most of the Pacific Ocean from Japan to New Guinea. At least once, *Discovery* passed directly over the eye of the monster storm.

At sunrise and sunset, layers of the atmosphere become evident as sheer colored bands. Instruments on the CRISTA-SPAS free-flyer often pointed to the horizon to measure ozone and other chemicals there.

STS-91: JUNE 2–12, 1998

9 : 19 : 54
DAYS HRS MINS

0 : 0
HRS MINS
0

235 M
204 NM
(378 KM)

154 | 51.6°

KENNEDY SPACE
CENTER

COMMANDER

- Charles J. Precourt, USAF, his last of 4 flights

PILOT

- Dominic L. Pudwill Gorie, USN, his 1st of 4 flights

MISSION SPECIALISTS

- Franklin R. Chang-Díaz, physicist, his 6th of 7 flights, 2 on *Discovery*

- Wendy B. Lawrence, USN, engineer, her 3rd of 4 flights, 2 on *Discovery*

- Janet L. Kavandi, chemist, her 1st of 3 flights

- Valery V. Ryumin, RSA, engineer, his only shuttle flight

- Andrew S. W. Thomas, mechanical engineer, his 2nd of 5 flights, 3 on *Discovery*, and a stay on Mir

The design for the last Shuttle-Mir mission depicts both spacecraft, both national flags, and both countries on the globe. The name in Cyrillic is cosmonaut Ryumin's.

From left to right: Dominic L. Pudwill Gorie and Charles J. Precourt with helmets, surrounded by Wendy B. Lawrence, Franklin R. Chang-Díaz, Janet L. Kavandi, Valery V. Ryumin, and Andrew S. W. Thomas.

Last Mission to Mir

Discovery flew twice in a row in 1998, and both missions made history. This one was the last of nine Shuttle-Mir docking missions and the final mission of the Shuttle-Mir program, yet it was *Discovery*'s only docking with Mir. *Discovery* had opened the Shuttle-Mir program with a rendezvous-only first visit to Mir (STS-63) in 1995 and thus bracketed the series of flights like a pair of bookends.

This last shuttle mission to Mir served several purposes. *Discovery* came to pick up astronaut Andy Thomas, who spent 130 days as the last American on a Mir mission. The shuttle crew also reclaimed U.S. experiments from Mir, bringing home equipment and samples from long months of materials science and life science research.

Discovery also brought about 6,000 pounds (2,722 kg) of water, food, and supplies to Mir to stock its current and future crews with extra clothing, batteries, equipment, new experiments, personal care packages, and more. The main crew activity during four days of docked operations was transferring these tons of baggage between the two spacecraft.

A SPACEHAB module flew on a Shuttle-Mir mission for the sixth time as a cargo container for most of the transferrable logistics items. However, it also had room for crew-tended experiments. Other mission science activities focused on middeck experiments and a solo instrument mounted behind SPACEHAB, the Alpha Mass Spectrometer. The prototype for one to be installed years later on the International Space Station, it was developed by a Nobel laureate in high-energy particle physics to study the universe via subatomic particles rather than wavelengths of light.

As the Shuttle-Mir missions came to an end, the American and Russian space agencies were ready for the next phase of cooperation: construction and operation of the International Space Station. The two partners had learned to share expertise and responsibility. Crews trained in both nations to integrate well into joint missions. Russia gained access to a huge cargo vehicle, and NASA gained access to an established space station. Seven U.S. astronauts spent a cumulative 907 days (2.5 years)

on longer missions than were possible on the shuttle, and experiments benefited from long months in orbit compared to one- to two-week shuttle flights that were briefer than desirable for microgravity research.

In technical preparation for International Space Station missions, two new pieces of equipment made their first appearance on STS-91. *Discovery* launched on a super lightweight external tank that was stronger and 7,500 pounds (3,400 kg) lighter than previous tanks; trimming the vehicle's weight increased payload-to-orbit capacity. *Discovery* also used the actual ISS docking mechanism for the first time instead of the prototype used for the other Shuttle-Mir dockings.

The Shuttle-Mir missions surfaced both the benefits and difficulties of an international endeavor in space. Although the road to the International Space Station had some bumps along the way, it was smoothed by the experience of these few years of working together. Astronauts and cosmonauts who came of age when the United States and the Soviet Union were Cold War adversaries modeled a new style of spaceflight: cooperation, not competition.

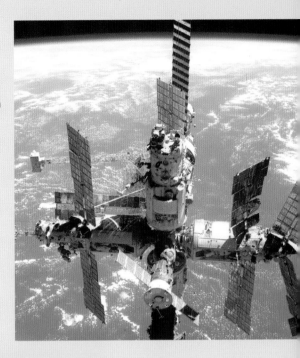

Above: Close-up photos of Mir from the shuttle showed how it had expanded module by module in various directions. They also revealed details of its condition, such as the damaged solar array to the right of center.

Left: This view from Mir shows the layout of *Discovery*'s payload bay, from the airlock and docking adapter (forward) to the crew tunnel and SPACEHAB module (mid-bay) and the separate square particle detector (aft).

Below: The crews for the last joint Shuttle-Mir mission pose in Mir. The Mir cosmonauts in blue are Talgat Musabayev (center) and Nikolai Budarin. Before donning the red shuttle crew shirt, Andrew S. W. Thomas (behind Musabayev) spent 130 days on Mir, and he logged 141 consecutive days in space from his launch to landing.

8 : 21 : 44
DAYS HRS MINS

349 M
303 NM
(561 KM)

134 | 28.5°

0 : 0
HRS MINS

0

KENNEDY SPACE CENTER

25TH *DISCOVERY* MISSION

COMMANDER

- Curtis L. Brown Jr., USAF, his 5th of 6 flights, 3 on *Discovery*

PILOT

- Steven W. Lindsey, USAF, his 2nd of 5 flights, 3 on *Discovery*

MISSION SPECIALISTS

- Stephen K. Robinson, mechanical engineer, his 2nd of 4 flights, 3 on *Discovery*
- Scott E. Parazynski, medical doctor, his 3rd of 5 flights, 2 on *Discovery*
- Pedro Duque, ESA, aeronautics engineer, his only flight

PAYLOAD SPECIALISTS

- Chiaki Mukai, JAXA, medical doctor, her 2nd and last flight
- John H. Glenn Jr., USMC, pilot and politician, his 2nd space flight; 1st flew on 1962 Mercury mission

A Hero's Return to Space

Discovery's twenty-fifth flight made many more headlines than usual, but not for reaching this milestone. Instead, an especially distinctive member of the crew captured media and public attention: a seventy-seven-year-old U.S. senator who had been the first American in orbit. Almost thirty-seven years after his 1962 Mercury flight, national hero John Glenn returned to space on the shuttle. He was the only original astronaut to fly on the shuttle.

STS-95 was a science mission, and *Discovery* carried more than eighty experiments in the middeck, SPACEHAB module, and payload bay. Experiments clustered in three primary fields: medical, materials, and solar research.

This was the twelfth flight of a pressurized SPACEHAB module, a compact commercial carrier that served as a second middeck. It could hold almost 5,000 pounds (2,268 kg) of crew-tended or automated experiments in racks or lockers sized to be interchangeable with middeck lockers. It also had a glovebox for working with specimens and chemicals.

Most medical and life science research in space addresses the effects of prolonged weightlessness on the human body. STS-95 investigations probed the crewmembers' balance, perception, immune system, bones and muscles, metabolism, blood flow, and sleep in microgravity to look for changes. Because some spaceflight adaptation effects are similar to the effects of aging, NASA justified John Glenn's presence on the flight as an elderly research subject. Although some thought that was a cover for rewarding a lifelong proponent of spaceflight with a shuttle trip, Glenn took his research role seriously.

Two other scientific payloads rode outside the lab module in the payload bay. SPARTAN 201 made its fourth flight, all but one on

Styled like the numeral 7 on John Glenn's *Friendship 7* capsule, this 7 signifies both the original seven Mercury astronauts and *Discovery*'s seven crewmembers. A tiny Mercury capsule orbits the shuttle, whose colorful exhaust plumes symbolize three fields of scientific research on this mission.

Front (left to right): Steven W. Lindsey and Curtis L. Brown Jr. Back (left to right): Scott E. Parazynski, Stephen K. Robinson, Chiaki Mukai, Pedro Duque, and John H. Glenn Jr.

Discovery, to study the sun. SPARTAN bore two complementary telescopes and spent two days as a free-flyer. A Hitchhiker mount held six instruments for astronomical observations in the extreme ultraviolet wavelength range, a signature of the birth and death of stars.

Another significant payload, called HOST, involved tests of hardware to be used on the third Hubble Space Telescope servicing mission (STS-103 flown by *Discovery* in 1999). Prior exposure of advanced technologies to the space environment built confidence that a new computer, solid state recorder, and cooling equipment would perform properly when installed on the telescope.

The STS-95 mission marked several other firsts. Payload Specialist Pedro Duque from the European Space Agency became the first person from Spain to fly in space. Chiaki Mukai, already the first Japanese woman in space, became the first person from Japan to fly more than once in space. This mission included the first high-definition television broadcast; Walter Cronkite and former astronaut Pete Conrad gave commentary for live national coverage of *Discovery*'s liftoff and John Glenn's return to space.

STS-95 is remembered primarily for John Glenn's return to space. Unusually large crowds attended the launch, including President Bill Clinton and First Lady Hillary Clinton. Glenn's star power as a space pioneer and admired national figure drew renewed attention to human spaceflight and briefly elevated public interest. But beyond the celebrity, STS-95 was a mission solidly dedicated to science. *Discovery* operated as a research facility in space, but on a scale soon to be eclipsed by the International Space Station.

Grasped by the robotic arm, SPARTAN is guided back to its berthing in the payload bay after two days of free-flight about 70 to 100 miles (113 to 161 km) from *Discovery*.

Above: President and First Lady Clinton witness the STS-95 launch with astronauts Robert D. Cabana and Eileen M. Collins and other NASA officials.

Left: An active participant in on-board science activities, Senator John Glenn wears equipment for sleep research. He is "standing" by his bunk in a cabinet of four middeck sleep compartments used on missions with two work shifts.

9 : 19 : 13
DAYS HRS MINS

153 | 51.6°

246 M
214 NM
(396 KM)

7 : 55
HRS MINS

1

KENNEDY SPACE CENTER

COMMANDER

- Kent V. Rominger, USN, his 4th of 5 flights, 2 on *Discovery*

PILOT

- Richard D. (Rick) Husband, USAF, his 1st of 2 flights

MISSION SPECIALISTS

- Tamara E. Jernigan, space physicist and astronomer, her last of 5 flights, 1 EVA

- Ellen L. Ochoa, electrical engineer, her 3rd of 4 flights, 2 on *Discovery*, RMS operator

- Daniel T. Barry, electrical engineer and medical doctor, his 2nd of 3 flights, 2 on *Discovery*, 1 EVA

- Julie Payette, CSA, engineer, her 1st of 2 flights

- Valery I. Tokarev, RSA, his only shuttle flight

Discovery and the International Space Station: A Story in Thirteen Episodes

Docking

Assembly of the International Space Station (ISS) spanned thirteen years and thirty-seven Space Shuttle missions. *Discovery* played a leading role while making the trip thirteen times. The ISS story should be told as a continuous narrative to reveal the methodical step-by-step buildup from a single module to a complex structure the size of a U.S. football field. But the episodes featuring *Discovery* are different enough to satisfy as a digest of the full program. *Discovery* flew the second mission to the ISS and also the second-to-last mission—framing the ISS story like bookends as it had also framed the Shuttle-Mir program.

The construction story began with the first space station element, the Russian Zarya ("Sunrise, Dawn") core power systems module with solar arrays, launched into orbit on a Proton rocket in November 1998. In December, on the first assembly mission, *Endeavour* brought the first U.S. element, the Unity connecting node, which the crew mated to Zarya. These two modules formed the nucleus from which the full space station expanded. In May 1999, on the first logistics mission to deliver supplies, *Discovery* made the first docking with the combined ISS elements, and the STS-96 crew entered to prepare the place for occupancy.

In-flight photos show members of the crew wielding screwdrivers and drills to mount interior hardware, running electrical cables, and generally doing finish work inside the space station. They also transferred about two tons of equipment for further interior outfitting as well as clothing, sleeping bags, computers, cameras, medical equipment, and drinking water for the first residents. Traffic flowed steadily from the

In this version of the astronaut symbol, the star represents the five participating space agencies, the colored rays honor the three national flags of the crewmembers, and the orbit circle is *Discovery*'s path to dock with the nascent space station.

Front (left to right): Kent V. Rominger, Ellen L. Ochoa, and Richard D. (Rick) Husband. Back (left to right): Daniel T. Barry, Julie Payette, Valery I. Tokarev, and Tamara E. Jernigan.

SPACEHAB double module in the aft half of *Discovery*'s payload bay where the supplies were packed, through the crew tunnel and docking adapter, and into the Unity node and Zarya.

Working outside for almost eight hours while Ellen Ochoa operated *Discovery*'s arm, Dan Barry and Tammy Jernigan installed the Russian Strela ("Arrow") cargo crane and a U.S. crane. They pre-positioned bags of tools and handrails for future assembly tasks, and they set up foot restraints for the next spacewalkers. The EVA sped along as they completed various tasks and inspections of exterior hardware.

Given the challenges of rendezvous, docking, and doing any work in weightlessness, it may be oversimplifying to characterize this mission as setting up a household, but that was the crew's primary goal—to get the International Space Station ready for its first residents.

On the STS-96 mission, *Discovery* transitioned from its recent flight history of scientific research to its final role as a heavy-duty cargo carrier. On its next mission (STS-103), *Discovery* was filled with equipment for servicing the Hubble Space Telescope, and then it flew twelve more times to the International Space Station, bearing huge structures or tons of supplies. In the early years of the new century, all of the orbiters finally became what they were destined to be: shuttles transporting people and equipment between Earth and an outpost in space.

Above: The STS-96 crew wrapped up their space station work by leaving some humorous signs in the Unity node to help the first residents get oriented.

Left: The sprawling International Space Station grew from this two-module nucleus, viewed from *Discovery* on the inspection fly-around before departure.

246 M
214 NM
(396 KM)

202 | 51.6°

12 : 21 : 42
DAYS   HRS   MINS

27 : 19
HRS   MINS

4

EDWARDS
AIR FORCE BASE

100TH SPACE SHUTTLE MISSION

COMMANDER

- Brian Duffy, USAF, his last of 4 flights

PILOT

- Pamela A. Melroy, USAF, her 1st of 3 flights, 2 on *Discovery*

MISSION SPECIALISTS

- Leroy Chiao, chemical engineer, his last of 3 shuttle flights, and an ISS expedition, 2 EVAs

- William S. McArthur Jr., USA, aerospace engineer, his last of 3 shuttle flights, and an ISS expedition, 2 EVAs

- Peter J. K. (Jeff) Wisoff, physicist, his last of 4 flights, 2 EVAs

- Michael E. López-Alegría, USN, aeronautics engineer, his 2nd of 3 shuttle flights, and an ISS expedition, 2 EVAs

- Koichi Wakata, NASDA, aerospace engineer, his 2nd of 4 shuttle flights, and two ISS expeditions, RMS operator

A vivid Z made from one ray of the astronaut corps symbol represents the truss added to the three-module space station, superimposed on the shuttle shape to show its increased size.

From left to right: Pamela A. Melroy, Leroy Chiao, Michael E. López-Alegría, William S. McArthur Jr., Peter J. K. (Jeff) Wisoff, Koichi Wakata, and Brian Duffy.

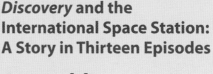

Discovery and the
**International Space Station:
A Story in Thirteen Episodes**

Assembly

After its STS-103 mission to revitalize the Hubble Space Telescope, *Discovery* entered full-time shuttle service to the International Space Station. On this, its second trip to the ISS, *Discovery* flew an assembly mission, bringing two critical elements for the station's continued expansion. Just a few days later, the first three residents moved in.

Since *Discovery*'s first visit here, Russia's Zvezda service module (living quarters) arrived in July 2000 and mated automatically with Zarya at the opposite end from Unity. Two other shuttle missions brought more supplies and equipment. Now it was this crew's job to deliver and install the first major exterior elements.

STS-92 was the one hundredth Space Shuttle mission, a milestone reached within twenty years and a distinctive mission in its own right. In four EVA days, the crew installed both the base for the huge solar array wings that would become a defining feature of the ISS and an extra docking port. Pam Melroy became the third female U.S. space pilot; she would soon become the second woman commander, also on a *Discovery* mission.

Installation of the zenith Z1 truss topped the list of mission tasks. The framework for the solar arrays would eventually be attached to this massive girder on top of the Unity node, but first it would support smaller solar arrays installed on the next shuttle mission. Using the manipulator arm, mission specialist Koichi Wakata attached the truss to the Unity node; then the EVA astronauts installed power converters, electrical lines, communications system antennas, and other equipment on it. The Z1 truss also housed gyroscopic devices for maintaining the ISS in proper orientation.

The crew's second major task was to install a third docking station on Unity. The circular node had six ports, four around and one on each end that accepted a pressurized mating adapter for docking a spacecraft or an additional module. Unity had two end adapters already, one attached to Zarya and the other used for shuttle dockings. The additional adapter gave the ISS an optional location for dockings and expansion.

While Wakata operated the arm, two spacewalking pairs completed these installations and other tasks. Leroy Chiao and Bill McArthur were responsible for the Z1 truss installation, and Jeff Wisoff and Mike López-Alegría handled the mating adapter work and also tested the SAFER rescue backpack while tethered to *Discovery*. Besides the rendezvous and departure maneuvers, Brian Duffy and Pam Melroy supported the installations with close-in maneuvering and gave ISS an altitude boost.

Crewmembers transferred more supplies and equipment from the orbiter into the station and made internal electrical connections with the truss. The coming resident crew would find the ISS ready for them to set up housekeeping. Over time, four of the five STS-92 mission specialists also would live and work on the space station they helped to build.

Below, left: This view from *Discovery*'s departure looks straight at the docking end of the round Unity node with its new features: the boxy Z1 truss structure and communications antenna on the right and the second conical mating adapter on the left. The Russian modules extend behind Unity, unseen except for a pair of solar array wings.

Below, right: The next shuttle crew approaching ISS from below could see some of the *Discovery* crew's work. Attached to the Unity node (top) are a conical black mating adapter on the nadir side and an antenna extending from the Z1 truss on the zenith side.

237 M
206 NM
(331 KM)

201 | 51.6°

2

KENNEDY SPACE
CENTER

COMMANDER

- James D. Wetherbee, USN, his 5th of 6 flights, 2 on *Discovery*

PILOT

- James M. Kelly, USAF, his 1st of 2 flights, both on *Discovery*

MISSION SPECIALISTS

- Andrew S. W. Thomas, mechanical engineer, his 3rd of 5 flights, 3 on *Discovery*, and a stay on Mir, 1 EVA
- Paul W. Richards, mechanical engineer, his only flight, 1 EVA

MISSION SPECIALISTS (UP)

- James S. Voss, U.S. Army, aerospace engineer, his 5th of 6 flights, 2 on *Discovery*, 1 EVA
- Susan J. Helms, USAF, aeronautics-astronautics engineer, her 1st of 5 flights, 2 on *Discovery*, 1 EVA
- Yury V. Usachev, RSA, engineer, his last of 2 shuttle flights

MISSION SPECIALISTS (DOWN)

- Sergei K. Krikalev, RSA, mechanical engineer, his last of 3 shuttle flights
- William M. Shepherd, USN, mechanical engineer, his last of 4 shuttle flights, and an ISS expedition
- Yuri P. Gidzenko, RSA, his only shuttle flight

The curved tri-color banner represents the exchange of three crewmembers on the first ISS crew rotation mission. Orbiter crew names are above the banner, and ISS crew names are below.

Discovery and the International Space Station: A Story in Thirteen Episodes

Crew Change

By the time *Discovery* returned to the ISS five months later, two big changes had occurred. One shuttle crew had installed the first port side truss and another had delivered and mated the U.S. Destiny laboratory module to the front side of the Unity node. The line of modules had grown longer, and the first perpendicular extension was in place.

Discovery's STS-102 mission had a new purpose: crew rotation. For the first time, a new ISS crew shuttled up to the space station and a departing crew shuttled home in the same vehicle. The first resident crew had traveled to the space station in a Russian Soyuz craft. From the arrival of this second crew until 2010, crews generally commuted to and from the ISS on the shuttle, and *Discovery* flew six crew rotation missions. For a decade, "up" and "down" appeared beside names on shuttle crew rosters.

STS-102 also was a logistics mission, introducing a new and larger cargo carrier called a Multipurpose Logistics Module (MPLM). Made in Italy, a member nation of the European Space Agency, the first of these reusable modules was named Leonardo. Over time, it was repeatedly loaded with supplies and equipment, flown on four

Discovery crew (center, from left): James M. Kelly, Andrew S. W. Thomas, James D. Wetherbee, and Paul W. Richards. ISS Expedition 1 crew (lower left, from left): Sergei K. Krikalev, William M. Shepherd, and Yuri P. Gidzenko. ISS Expedition 2 crew (lower right, from left): James S. Voss, Yury V. Usachev, and Susan J. Helms.

more *Discovery* missions (and three times on other orbiters), and finally left permanently attached to the ISS on *Discovery*'s last mission.

Instead of transferring all goods from the shuttle to the space station one package at a time, as on previous logistics missions that had used the SPACEHAB cargo container, Leonardo itself was transferred to the ISS. The remote manipulator arm operator lifted it out of the payload bay, oriented it in the right direction, and moved it to a docking station on the Unity node. Once it was securely attached, the ISS crew opened the hatches and more efficiently transported cargo into the station.

The *Discovery* crew carried out an EVA to clear the path for Leonardo by relocating a pressurized mating adapter and rearranging some other hardware. They also installed on Destiny's outer wall a support base for a soon-to-be-delivered Canadarm similar to the one in the orbiter payload bay. Upon completing their work, Jim Voss and Susan Helms had made the longest EVA in shuttle history: 8 hours and 56 minutes. Andy Thomas and Paul Richards did another 6-hour EVA to install a spare parts platform and prepare the robotic arm base.

Meanwhile, the indoor crew began to outfit the U.S. Destiny lab with the first scientific rack—the Human Research Facility—and the robotic arm workstation. After unloading and stowing about 5 tons of cargo and 1,000 pounds of drinking water, they reloaded Leonardo with about 1,600 pounds of items and trash for return to Earth. Then arm operator Thomas retrieved the cargo container and placed it back in the payload bay.

In the transition from ISS Expedition 1 to Expedition 2, *Discovery*'s mission again demonstrated the shuttle's versatility, and it became the pattern for many successive missions: transport a fresh crew for an ISS change of command, deliver a very large container filled with whatever was needed, continue outfitting the interior and exterior of the station, and return the previous crew. *Discovery*'s next mission moved to a similar rhythm.

Above left: Astronaut Susan J. Helms, anchored to the end of *Discovery*'s arm, moves equipment to be installed on the space station.

Above right: Cosmonaut Yuri P. Gidzenko works inside the Leonardo module while it is attached to the ISS. The combined shuttle and ISS crews moved large equipment racks and tons of smaller equipment, supplies, and water packed in the large white bags.

STS-105: AUGUST 10–22, 2001

252 M
219NM
(406 KM)

186 | 51.6°

2

11 : 45
HRS   MINS

KENNEDY SPACE CENTER

30TH *DISCOVERY* MISSION

COMMANDER

- Scott J. Horowitz, USAF, his last of 4 flights, 2 on *Discovery*, RMS operator

PILOT

- Frederick W. (Rick) Sturckow, USMC, his 2nd of 4 flights, 2 on *Discovery*

MISSION SPECIALISTS

- Patrick G. Forrester, USA, engineer, his 1st of 3 flights, 2 on *Discovery*, 2 EVAs

- Daniel T. Barry, medical doctor and electrical engineer, his last of 3 flights, 2 on *Discovery*, 2 EVAs

ISS CREW (UP)

- Frank L. Culbertson Jr., USN, engineer, his 3rd of 4 flights

- Vladimir N. Dezhurov, RSA, engineer, his 1st of 2 shuttle flights

- Mikhail Tyurin, RSA, engineer, his 1st of 2 shuttle flights

ISS CREW (DOWN)

- James S. Voss, USA, engineer, his 2nd of 4 flights, 2 on *Discovery*

- Susan J. Helms, USAF, engineer, her last of 6 flights, 3 on *Discovery*

- Yury V. Usachev, RSA, engineer, his 2nd of 2 shuttle flights

The curved flag banners and gold stars represent the ISS crew rotation cycle, with three stars for the ascending U.S.-commanded Expedition 3 crew and two stars for the descending Russia-commanded Expedition 2 crew.

Discovery and the International Space Station: A Story in Thirteen Episodes

Logistics

Four months later, *Discovery* headed into space for the second time in 2001. Its thirtieth mission was similar to its most recent one to bring a crew and supplies to the International Space Station. The Leonardo module flew for the second time, and the second ISS crew rotation occurred.

Two assembly missions between these two *Discovery* flights added new features to the station: a long Canadarm to move spacewalkers and equipment into position and an airlock for extravehicular activities. Now the space station had EVA capability comparable to the shuttle and another portal for going to work outside whether or not the shuttle was present.

On this trip, *Discovery* delivered the third ISS crew and picked up the second crew, who had flown up on this vehicle, to bring them home. The Expedition 2 crew spent five months on the space station continuing to configure it for full operations. *Discovery* was the first orbiter to provide roundtrip service for a single ISS crew, now effectively becoming a commuter vehicle for astronauts assigned to station duty.

As before, Leonardo came filled with several tons of supplies and equipment. Once it was berthed to the Unity node and activated, the

ISS Expedition 2 crew (left): James S. Voss, Yury V. Usachev, and Susan J. Helms. ISS Expedition 3 crew (right): Mikhail Tyurin, Frank L. Culbertson, Jr., and Vladimir N. Dezhurov. Orbiter crew (center): Frederick W. (Rick) Sturckow, Patrick G. Forrester, Daniel T. Barry, and Scott J. Horowitz.

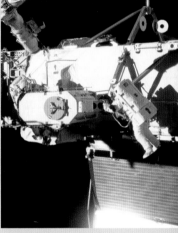

combined crews set to work unloading. A designated loadmaster—Dan Barry on this mission—always orchestrated these transfers to ensure that supplies and equipment migrated in priority order and followed a detailed stowage plan so the crew could easily find everything as they unpacked.

Among the most welcome cargo items were two large racks of scientific equipment for the U.S. Destiny laboratory module. When fully equipped with twenty-four such racks, Destiny was to be the prime location for American research activities on the ISS. NASA's slogan for the mission, "Science Moves to the Forefront," suggested the priority of starting scientific work while basic construction continued.

Two EVAs highlighted this mission, first to install an ammonia servicer unit for the station's cooling systems and attach an experiment, and then to install EVA handrails and cables for another truss. The experiment exposed a variety of materials directly to the space environment to learn how they weathered. Dan Barry and Pat Forrester performed both EVAs, supported by Scott Horowitz and Rick Sturckow on the aft flight deck.

After completing this mission, *Discovery* rotated out of service for its third major modification period, becoming the first orbiter to undergo this work at Kennedy Space Center rather than in Palmdale, California. This time, *Discovery* received the glass cockpit upgrade and a thorough inspection and overhaul, resulting in more than 350 other changes. Meanwhile, *Atlantis* and *Endeavour* continued flying space station missions and *Columbia* flew the last dedicated science mission.

Above: Daniel T. Barry (left) and Patrick G. Forrester (center) install the ammonia servicer package on the ISS P6 truss after receiving it from the Canadarm. The end of the arm hovering nearby is visible above Barry after it released the grapple fixture on top of the unit.

Above left: Discovery approaches the ISS carrying the Leonardo logistics module and a payload carrier; the airlock-docking adapter occupies the forward payload bay.

STS-114: JULY 26–AUGUST 9, 2005

STS-121: JULY 4–17, 2006

STS-114

13 : 21 : 33
DAYS HRS MINS

220 M
191 NM
(354 KM)

219 | 51.6°

20 : 5
HRS MINS
3

EDWARDS
AIR FORCE BASE

COMMANDER

- Eileen M. Collins, USAF, her last of 4 flights, 2 on *Discovery*

PILOT

- James M. Kelly, USAF, his last of 2 flights, both on *Discovery*, RMS operator

MISSION SPECIALISTS

- Soichi Noguchi, JAXA, aeronautical engineer, his only shuttle flight, and an ISS expedition, 3 EVAs

- Stephen K. Robinson, mechanical engineer, his 3rd of 4 flights, 3 on *Discovery*, 3 EVAs

- Andrew S. W. Thomas, mechanical engineer, his last of 4 flights, 3 on *Discovery*

- Wendy B. Lawrence, USN, ocean engineer, her last of 4 flights, 2 on *Discovery*, RMS operator

- Charles J. Camarda, aerospace engineer, his only flight

The Space Shuttle returns to flight and honors the memory of the STS-107 *Columbia* crew with seven stars in the constellation Columba, also in their mission emblem's design. The blue ring representing the ISS orbit bears the names of EVA crewmembers.

Discovery and the International Space Station: A Story in Thirteen Episodes

Return to Flight

Discovery's intended eighteen months on the ground for major maintenance became a four-year hiatus after the unthinkable happened again: *Columbia* disintegrated and the seven STS-107 crewmembers perished during their return on February 1, 2003. The cause of the second shuttle tragedy was traced to a debris impact, a large piece of external tank insulation that broke off during ascent and fatally damaged one of the orbiter's wings. NASA halted shuttle flights until this and related problems were corrected. As in 1988, after a lengthy accident investigation that kept the shuttle fleet grounded, *Discovery*, freshly inspected and upgraded, became the return-to-flight orbiter.

The STS-114 and STS-121 test flight missions focused on evaluating new safety measures. Apart from changes in the external tank to reduce debris, the orbiter received several modifications to monitor its condition better. New cameras were installed on the underside of the vehicle, the solid rocket boosters, and the tank to record falling and

In orange (from left): James M. Kelly, Andrew S. W. Thomas, Wendy B. Lawrence, Charles J. Camarda, and Eileen M. Collins. In white (from left): Stephen K. Robinson and Soichi Noguchi.

Most of *Discovery*'s underside is visible in this photograph, which was made by an ISS Expedition 13 crewmember during the orbiter's "backflip" maneuver before docking.

possibly damaging insulation. NASA adopted a policy of daylight-only launches for monitoring the tank and recording any debris strikes. Teams on the ground scrutinized images from at least one hundred cameras to determine whether the orbiter sustained damage on its way to orbit.

STS-121 became a second return-to-flight mission because cameras showed tank insulation still falling off during the STS-114 launch. Although *Discovery* sustained no serious damage, this debris shedding remained a concern; NASA decided not to resume flights until more work resolved the problem. STS-121 might have flown sooner, but Hurricane Katrina roared through the Gulf of Mexico, damaging NASA's external tank production plant in Louisiana and interrupting the redesign work.

New equipment and in-flight inspection techniques were developed for *Discovery*'s crews to test. The orbiter gained a second 50-foot (15-meter) arm, a straight one tipped with a sensor package of cameras and a laser device that attached to the jointed Canadarm in orbit. The arm operators used the Orbiter Boom Sensor System (OBSS) extension to scan every inch of the orbiter—especially the previously inaccessible nose, back, and underside—in search of damage. In addition, the wing leading edges made of a reinforced

Seen through an orbiter window, the straight boom—tipped with a sensor package (far left)—doubles the reach of *Discovery*'s jointed remote manipulator arm.

carbon material were outfitted with impact detectors; several of those panels had been damaged on *Columbia*. Crews did these in-orbit inspections twice on all post-*Columbia* missions, early and late in the mission, to ensure that the orbiters were sound for reentry.

Mission commanders Eileen Collins and Steve Lindsey performed a new flight maneuver as *Discovery* approached the International Space Station, technically an R-bar pitch maneuver but more casually called a rendezvous pitch maneuver or "backflip." They and all subsequent commanders (except on the STS-125 Hubble servicing mission) brought the orbiter nose-first to a position 600 feet (183 meters) from the station. Then they slowly pitched it up and over to reveal the entire underside and aft end to ISS crewmembers armed with cameras at a window. Their digital images of the protective tiles and leading edge panels were transmitted to the ground for

STS-121

12 : 18 : 38
DAYS HRS MINS

202 | 51.6°

219 M
190 NM
(352 KM)

21 : 29
HRS MINS

3

KENNEDY SPACE
CENTER

115TH SPACE SHUTTLE MISSION

COMMANDER:

- Steven W. Lindsey, USAF, his 4th of 5 flights,
  3 on *Discovery*

PILOT:

- Mark E. Kelly, USN, his 3rd of 4 flights, 2 on *Discovery*

MISSION SPECIALISTS:

- Michael E. Fossum, systems engineer, his 1st of 3 flights,
  2 on *Discovery*, 3 EVAs

- Lisa M. Nowak, USN, aeronautics-astronautics engineer,
  her only flight, RMS operator

- Stephanie D. Wilson, aerospace engineer, her 1st of
  3 flights, all on *Discovery*, RMS operator

- Piers J. Sellers, biometeorologist, his 2nd of 3 flights, 3 EVAs

ISS CREW (UP):

- Thomas A. Reiter, ESA, aerospace engineer, roundtrip
  shuttle flights for his ISS expedition, both on *Discovery*

Superimposed on
the symbol of the
Astronaut Office, the
shuttle is docked at
the space station.

From left: Stephanie D. Wilson, Michael E. Fossum, Steven W. Lindsey, Piers J. Sellers,
Mark E. Kelly, Thomas A. Reiter, and Lisa M. Nowak.

Wendy B. Lawrence and James M. Kelly operate the space station's Canadarm at the
control station in the Destiny lab. Without a window, they rely on laptop computers
for data and views from cameras, plus comments from the EVA crew.

examination. The ten-minute maneuver gave added insurance that no
damage went undetected.

The crews also tested tools and techniques for in-orbit tile and
leading edge repairs. Two of three STS-114 EVAs included trial repairs.
The first occurred at a workstation in the payload bay with samples
of damaged tiles and carbon panels; the astronauts applied different
adhesives to cracks and gouges using tools that resembled caulk guns
and paint scrapers. An unexpected task was added to the third EVA after
inspection showed two protruding tile gap fillers below *Discovery*'s nose.
Steve Robinson, mounted on the station's arm, performed the first EVA
underneath an orbiter to pluck these strips out so they would not disrupt
airflow and overheat the vehicle during reentry. The STS-121 EVA team
also tried repairing damaged sample tiles and carbon panels at a payload
bay workstation. Various repair methods proved feasible, and repair
materials were available on all later missions in case of need.

Besides the primary return-to-flight objectives, these two missions also met ISS logistics needs. The crews brought much-needed supplies and equipment to the space station and also did some installation and maintenance tasks. *Discovery* brought the Raffaello logistics module on STS-114, and its twin Leonardo came on STS-121, each loaded with tons of necessities. The STS-114 EVA team replaced one of the station's attitude control gyro units and installed an equipment storage platform and a large experiment on the station exterior. The STS-121 EVA team repaired the mobile transporter for the station's Canadarm, delivered a spare pump for the ISS thermal control system, and tested the stability of the OBSS boom as an EVA work station. These EVAs involved close coordination with one or both of the Canadarms and their operators.

The STS-121 mission also transported European astronaut Thomas Reiter to bring the ISS crew complement back to three. While the fleet was grounded, the small, three-person Soyuz capsule became the sole crew vehicle. The ISS crew shrank from three to two residents, all that could be supported without provisioning by the shuttle, and the stays stretched longer to simplify ferry operations. Now the shuttle was back in business for crew rotations and the planned expansion of the ISS crew to six people when more modules were in place.

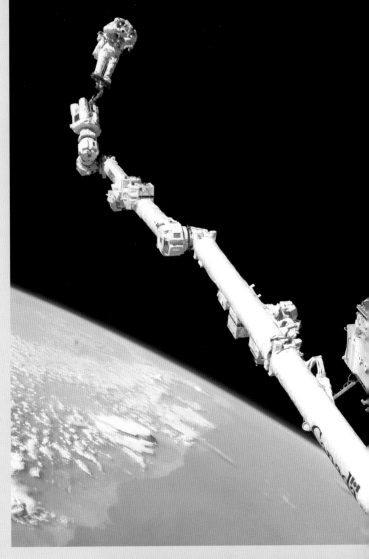

STS-121 was the 115th shuttle mission; nineteen more followed without incident after regular flights resumed in 2007. Smaller and fewer bits of tank insulation continued to fall but did not damage the orbiters. The stringent new safety measures adopted after the loss of *Columbia* gave the entire mission team much better information about the orbiter's condition and gave the crew several options for repairing damage in orbit. For a worst-case scenario of grave damage, NASA developed a contingency plan for the shuttle crew to board the space station until a rescue mission could be launched. From STS-114 on, another orbiter was always available to launch on short notice, except for the last mission, STS-135. Fortunately, no such emergencies arose.

The return-to-flight missions accomplished what they were supposed to do: restore confidence in the shuttle to resume missions and complete the space station. Yet the loss of *Columbia* was a fresh reminder that human spaceflight remains experimental, and it provoked a reassessment of the Space Shuttle's future. By the time *Discovery* flew these two missions, an end date for the shuttle program had been announced and NASA was already beginning to plan for a new vehicle. But *Discovery* had seven missions yet to fly.

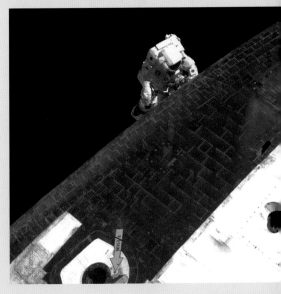

Above and right: Stephen K. Robinson, riding on the space station's Canadarm, makes the first EVA under an orbiter. He ventures there to remove two small pieces of material that are protruding between tiles.

STS-116: DECEMBER 9–22, 2006

12 : 20 : 45
DAYS   HRS   MINS

213 M
185 NM
(343 KM)

204 | 51.6°

25 : 45
HRS   MINS

4

KENNEDY SPACE
CENTER

COMMANDER

- Mark L. Polansky, USAF, his 2nd of 3 flights

PILOT

- William A. Oefelein, USN, his only flight

MISSION SPECIALISTS

- Nicholas J. M. Patrick, mechanical engineer, his 1st of 2 flights, RMS operator

- Robert L. Curbeam Jr., USN, aeronautics-astronautics engineer, his last of 3 flights, 2 on *Discovery*, 4 EVAs

- Christer Fuglesang, ESA, physicist, his 1st of 2 flights, both on *Discovery*, 3 EVAs

- Joan E. Higginbotham, electrical engineer, her only flight

ISS CREW (UP)

- Sunita L. Williams, USN, her 1st of 2 shuttle flights and 2 ISS expeditions, 1 EVA

ISS CREW (DOWN)

- Thomas Reiter, ESA, aerospace engineer, roundtrip shuttle flights for his ISS expedition, both on *Discovery*, and a mission to Mir

Discovery and the
International Space Station:
A Story in Thirteen Episodes

Power

After the two return-to-flight missions, *Discovery* entered a fairly regular rotation with *Atlantis* and *Endeavour* for the last five years of the Space Shuttle program. On this assembly mission, the twentieth shuttle flight to the International Space Station and one of the most complex missions in shuttle history, *Discovery*'s crew reconfigured a truss and solar arrays for an expanding power system. At a pivotal moment, the space station's permanent power system switched on.

NASA's theme for STS-116 was "Power Up for Science." In the near future, European and Japanese laboratory modules would be added to the ISS, and the station's power plant had to supply the increased demand for electricity. *Discovery* brought a port-side truss segment (P5) to continue expansion of the station's solar power framework, and astronauts used the orbiter's arm to transfer the segment to the space station's arm for positioning. The crew spent three EVA days installing the truss section and its many power, data, and heater cables and completely rewiring the station for future capacity.

The crew also retracted a solar array so another recently installed array had room to rotate and track the sun. That was when they met a problem. The older array did not fold properly into flat pleats like an accordion, but some sections kinked or bulged, jamming the process and leaving the array partly extended. Electronic commands and manual efforts, even an attempt to jostle around inside the station to wiggle the array, did not work to make it fold. Finally, an extra EVA attempt

The North Star marks the position of the P5 truss between the space station's port-side solar arrays. The flag of Sweden recognizes the first flight of a Swedish astronaut on this crew.

Front (from left): William A. Oefelein, Joan E. Higginbotham, and Mark L. Polansky. Back (from left): Robert L. Curbeam Jr., Nicholas J. M. Patrick, Sunita L. Williams, and Christer Fuglesang.

succeeded in freeing and fully retracting the delicate array. During this effort, mission specialist Robert Curbeam became the first person to do four spacewalks in a single mission.

Discovery carried SPACEHAB filled with about two tons of supplies and equipment for the ISS and returned with an equal amount of stuff no longer needed there. One new ISS crewmember, Suni Williams, rode up to take the place of ESA astronaut Thomas Reiter, who returned on Discovery. After spending eight days docked to the station, the crew released three very small satellites for an experiment before heading home—the first such deployment since 2002.

Among other distinctions, the STS-116 crew was the first to include two African Americans and also the first Scandinavian astronaut, and it was one of only a few to include two Europeans. Five of this mission's crewmembers were rookies, as NASA aimed for all members of the astronaut corps to gain spaceflight experience on the shuttle.

NASA was planning to retire the Space Shuttles in 2010, but much work remained to complete the ISS first. With an enhanced power system now functioning, the International Space Station was ready to expand.

Robert L. Curbeam Jr. (left) and Christer Fuglesang install a new segment of the massive port-side truss on the first EVA.

15 : 2 : 24
DAYS HRS MINS

238 | 51.6°

216 M
188 NM
(348 KM)

27 : 14
HRS MINS

4

KENNEDY SPACE
CENTER

120TH SPACE SHUTTLE MISSION

COMMANDER

- Pamela A. Melroy, USAF, her last of 3 flights, 2 on *Discovery*

PILOT

- George D. Zamka, USMC, his 1st of 2 flights

MISSION SPECIALISTS

- Scott E. Parazynski, medical doctor, his last of 5 flights, 2 on *Discovery*, 4 EVAs

- Stephanie D. Wilson, aerospace engineer, her 2nd of 3 flights, all on *Discovery*, RMS operator

- Douglas H. Wheelock, USA, aerospace engineer, his only shuttle flight, and an ISS expedition, 3 EVAs

- Paolo A. Nespoli, ESA, aeronautics-astronautics engineer, his only shuttle flight, and an ISS expedition, RMS operator

ISS CREW (UP)

- Daniel M. Tani, mechanical engineer, his 2nd of 3 shuttle flights, 1 EVA

ISS CREW (DOWN)

- Clayton C. Anderson, aerospace engineer, his 2nd of 3 shuttle flights, 2 on *Discovery*

Beside the shuttle bearing the Harmony node, a bright star symbolizes the space station and the relocation of the P6 solar array truss from red points to gold points. Opposite are future destinations—the moon and Mars.

Discovery and the International Space Station: A Story in Thirteen Episodes

Harmony

NASA's slogan for the 120th Space Shuttle mission was "Harmony: A Global Gateway." *Discovery* carried the essential element—a multi-port node named Harmony—for adding laboratory modules prepared by Europe and Japan. Harmony was the first new living space added since 2001. This second connecting node would permit expansion of the International Space Station to its full research capacity and, when the labs were in place, a six-person crew.

    Discovery's second consecutive assembly mission included four heavy-duty extravehicular activities to reconfigure the space station. After STS-116, *Atlantis* and *Endeavour* assembly mission crews had built out the starboard side truss and its first pair of 112-foot-long (34 meters) solar array wings. *Discovery*'s crew delivered the new node and also relocated a second pair of solar wings to the port side, significantly changing the look of both space station axes—the line of modules and the perpendicular truss line.

    The EVAs unfolded like a ballet in four acts that featured a spacewalking duo with an ensemble of robotic arm operators, all interacting under the watchful eye of a director and stage manager. The first EVA focused on lifting Harmony out of the payload bay with the space station's arm and mating it temporarily to the port side of the Unity node while *Discovery* was docked to the end of the U.S. Destiny laboratory. The next flight day, the action continued inside as the crew

From left: Scott E. Parazynski, Douglas H. Wheelock, Stephanie D. Wilson, George D. Zamka, Pamela A. Melroy, Daniel M. Tani, and Paolo A. Nespoli. Clayton C. Anderson is not pictured.

Above: Discovery approaches the International Space Station to deliver the multi-port Harmony node.
Below: Before *Discovery* arrived, the ISS sported one pair each of starboard and port solar array wings. The second port pair, stowed in the four narrow, gray containers on their truss segment at the center of this image, was relocated and unfurled during this mission.

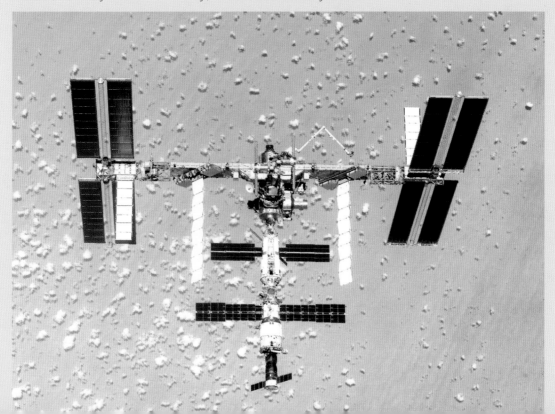

connected power and data cables and opened the hatch to ventilate the node so they could enter and start outfitting it. After *Discovery*'s departure, the ISS crew would use the station's arm again to reposition Harmony at the end of the Destiny module.

The second EVA focused on detaching the P6 solar array truss that rose above the Z1 truss; it had been in that temporary "vertical" location since *Discovery*'s STS-92 crew installed it there in 2000. The crew of *Discovery*'s previous STS-116 mission had already stowed the long arrays in preparation for this move. The EVA team disconnected it, and the inside team picked it up with the station's arm. On the next workday, they transferred it to the shuttle's arm, freeing the station's arm to move to the new installation site.

The crewmembers performed their parts deftly for the third EVA, successfully moving and installing the P6 solar array truss in its permanent position. Then the drama began. As the arrays unfolded, the crew noticed a rip and halted the process to avoid further damage. Two days of analysis and planning ensued on the ground. Then the crew spent a full day learning the repair procedure, assembling and making tools for it, and rehearsing for an urgent fourth EVA to repair and extend the solar array.

For that effort, the long straight boom from the orbiter was attached to the space station arm and the combo was fully extended to position Scott Parazynski beside the damaged wing. He made the repairs, with Doug Wheelock nearby coaching him and the arm operators through the process. After several suspenseful hours of concentration, the crew tried to deploy the array. Slowly and steadily it unfolded, to everyone's relief. This fine teamwork and recovery restored the ISS power system to full capacity.

Assembly tasks dominated the mission and made it *Discovery*'s second longest flight, but STS-120 also marked two quite different historic events. With Pam Melroy as shuttle commander and Peggy Whitson as ISS Expedition 16 commander, this was the first, and thus far only, time two women were simultaneously in command of space missions. In the fantasy realm, for the thirtieth anniversary of the *Star Wars* franchise, Luke Skywalker's lightsaber flew on *Discovery* as a special payload.

STS-120 was another pivotal mission in the space station's evolution. When *Discovery* departed after being docked for eleven days, the International Space Station looked much different than when the shuttle arrived. It was now almost ready for the arrival of European and Japanese lab modules, to be attached at right angles to the Harmony node on the next two shuttle missions. The ISS would then increase in habitable volume by about 20 percent to accommodate a larger crew and more research—a big advance toward its completion.

Berthing ports in the pristine Harmony node, newly attached to the International Space Station, await the coming Japanese Experiment Module (JEM, left), the European Columbus laboratory module (right), and the Multipurpose Logistics Module (MPLM). The shuttle would dock behind the hatch.

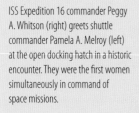

ISS Expedition 16 commander Peggy A. Whitson (right) greets shuttle commander Pamela A. Melroy (left) at the open docking hatch in a historic encounter. They were the first women simultaneously in command of space missions.

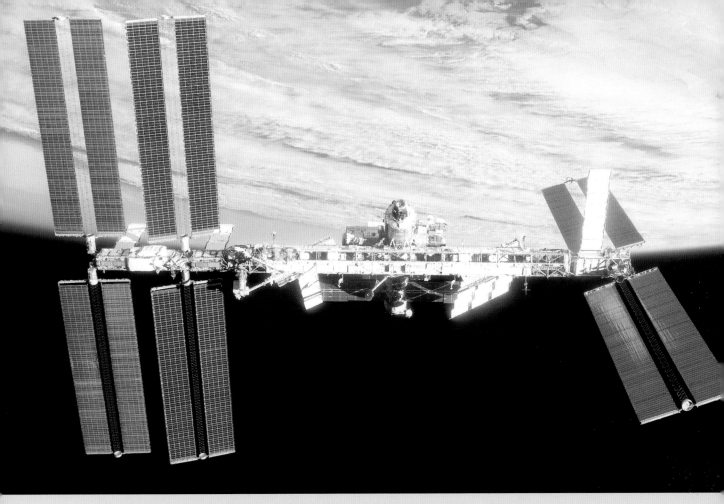

In this view from *Discovery*'s departure fly-around, the repositioned port solar arrays are to the far left and the cylindrical Harmony node is in its temporary sidelong position at the center of the image.

Mounted at the end of the fully extended, 100-foot ISS arm-orbiter boom combination, Scott E. Parazynski inspects the repaired solar array as it deploys. The rip is still visible near the center of the image, as are the strengthening wire loops that he threaded across the damaged area.

13 : 18 : 13
DAYS HRS MINS

219 M
190 NM
(352 KM)

217 | 51.6°

20 : 32
HRS MINS

3

KENNEDY SPACE
CENTER

35TH *DISCOVERY* MISSION

COMMANDER

• Mark E. Kelly, USN, his 3rd of 4 flights, 2 on *Discovery*

PILOT

• Kenneth T. Ham, USN, his 1st of 2 flights

MISSION SPECIALISTS

• Karen L. Nyberg, mechanical engineer, her only shuttle flight, and an ISS expedition, RMS operator

• Ronald J. Garan Jr., USAF, aerospace engineer, his only shuttle flight, and an ISS expedition, 3 EVAs

• Michael E. Fossum, USAF, systems engineer, his 2nd shuttle flight, both on *Discovery*, and an ISS expedition, 3 EVAs

• Akihiko Hoshide, JAXA, aerospace engineer, his only shuttle flight, and an ISS expedition, RMS operator

ISS CREW (UP)

• Gregory E. Chamitoff, aeronautic-astronautic engineer, his 1st of 3 shuttle flights

ISS CREW (DOWN)

• Garrett E. Reisman, mechanical engineer, his 1st of 3 shuttle flights

The flag of Japan and Japanese characters acknowledge the Kibō laboratory module delivered and installed on the space station by this mission crew.

Discovery and the International Space Station: A Story in Thirteen Episodes

Hope

On its thirty-fifth mission, *Discovery* delivered the largest space station element it ever carried and the largest ISS module—the pressurized laboratory for the Japanese Experiment Module (JEM) complex. This suite of research facilities arrived and was assembled over the course of three shuttle missions. Japan named its wing of the International Space Station Kibō, or Hope, and so NASA's theme for STS-124 was "Hope for a New Era." Europe's Columbus laboratory module had arrived three months earlier; with Kibō now in place, the era of more ambitious scientific research on the station could begin.

Part of Japan's facility, a logistics module filled with equipment, was already waiting there, temporarily attached to the Harmony node. This *Discovery* crew's priorities were to remove the 37-foot-long (11-meter) JEM lab module from the payload bay, attach it to the Harmony node across from the Columbus lab, and then relocate the logistics module to a port on the JEM lab. There it would serve as a storage room for experiments and spare equipment. *Discovery* also brought a 33-foot (10-meter) robotic arm to be installed on Japan's lab module.

This and other work consumed three EVA periods during *Discovery*'s eight days at the station. Spacewalkers Mike Fossum and Ron Garan performed the EVA tasks, and Karen Nyberg and Aki Hoshide operated

From left: Gregory E. Chamitoff, Michael E. Fossum, Kenneth T. Ham, Mark E. Kelly, Karen L. Nyberg, Ronald J. Garan Jr., and Akihiko Hoshide. Garrett E. Reisman is not pictured.

In this view from *Discovery*'s departure, the JEM Kibō elements are clearly visible where the central axis of modules extends to the right like an *L*. The long lab module sprouts an arm at the far end, and the round logistics module is perched on top of the lab.

the robotic arms, using the ISS arm to transfer modules. They also tested the six-jointed Kibō arm that would later move experiments to and from a platform outside the JEM lab. Meanwhile, Greg Chamitoff replaced Garrett Reisman on the ISS crew, and all hands began setting up equipment to activate the new lab and deal with another priority: repair of the station's toilet.

    Discovery's crew had an atypical task: retrieve an orbiter boom sensor system from the space station and reinstall it in the orbiter. The Kibō lab module was so large that the 50-foot (15-meter) boom would not fit in the payload bay for the trip up, yet it was needed for the external inspection of the orbiter to clear it for a safe return. To resolve this dilemma, the previous *Endeavour* crew left their OBSS boom parked on the ISS for *Discovery*'s use. One EVA was largely devoted to the boom's installation in the orbiter, and the inspection occurred near the end of the mission instead of the usual early timing.

    Other STS-124 novelties included the presence of the first European automated transfer vehicle (the Jules Verne ATV), already docked to the ISS on a supply run, and the presence of Buzz Lightyear, an action figure based on the animated character in the *Toy Story* films. Disney and Pixar teamed with NASA to produce educational programs based on Buzz Lightyear and space station science. *Discovery* provided the toy's ride up and returned it on a later mission.

    By the end of this *Discovery* mission, the Kibō lab and logistics modules were situated for ISS crews to embark on a much-expanded program of microgravity research in materials and life sciences. Japan's hopes for its space research program had reached orbit.

Although it is not yet furnished with racks of experiments, the Kibō laboratory module is an impressively spacious research facility.

217 M
189 NM
(350 KM)

202 | 51.6°

3

KENNEDY SPACE CENTER

125TH SPACE SHUTTLE MISSION

COMMANDER

- Lee J. Archambault, USAF, his 2nd of 2 flights

PILOT

- Dominic A. (Tony) Antonelli, USN, his 1st of 2 flights, RMS operator

MISSION SPECIALISTS

- Joseph M. Acaba, hydrogeologist and educator, his only shuttle flight, and an ISS expedition, 2 EVAs

- Steven R. Swanson, computer scientist-engineer, his 2nd of 2 flights, 2 EVAs

- Richard R. Arnold II, environmental scientist-educator, his only flight, 2 EVAs

- John L. Phillips, USN, physicist, his 2nd of 2 shuttle flights, and an ISS expedition, RMS operator

ISS CREW (UP)

- Koichi Wakata, JAXA, structural engineer, his 2nd of 3 shuttle flights, 2 on *Discovery*

ISS CREW (DOWN)

- Sandra H. Magnus, engineer, her 3rd of 4 flights

Shaped and colored to look like a solar array, the design highlights in gold the last segment of the ISS truss and solar wings. The numbers are the mission's dual designation as STS-119 and ISS-15A. Seventeen white stars commemorate the Apollo 1, *Challenger*, and *Columbia* astronauts.

Discovery and the International Space Station: A Story in Thirteen Episodes

Full Power

The first mission of 2009 was the 125th of the Space Shuttle era, and it was *Discovery*'s turn to fly. A *Discovery* crew also had flown on the one hundredth shuttle mission in 2000, and coincidentally STS-119 was the one hundredth mission since the *Challenger* accident. This was another ISS assembly mission with a partial crew change, and *Discovery*'s load was the last solar array truss segment. The crew had an enviable job: completing the ISS power grid.

A year and a half earlier, the STS-120 crew extended the port side truss by relocating the P6 solar array truss segment. The current crew turned its attention to the starboard side, delivering and installing the S6 segment. Its long radiator panel and two large solar array wings completed the ISS power system provided by the United States to produce 120 kilowatts of electricity for the lab modules and nodes.

Four spacewalkers and four manipulator arm operators collaborated to complete three extravehicular activities. The highest-priority tasks were to move the truss segment into position, bolt it in place, connect power and data cables, and unfurl the solar arrays—all done without a hitch. Also on the agenda: adding a GPS antenna to the Kibō laboratory module,

Clockwise from the mission emblem: Dominic A. (Tony) Antonelli, Joseph M. Acaba, John L. Phillips, Steven R. Swanson, Richard R. Arnold II, Koichi Wakata, and Lee J. Archambault. Sandra H. Magnus is not pictured.

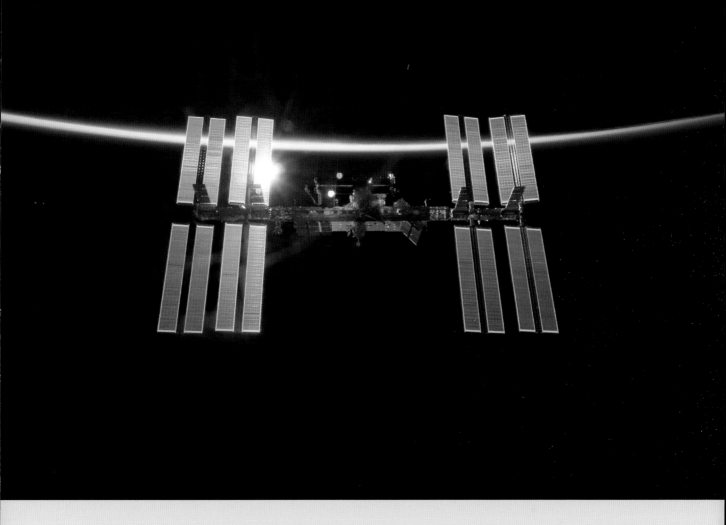

relocating equipment carts, preparing for future battery replacements, doing a variety of maintenance work on the ISS arm and truss, installing a wireless video system, and more.

Indoor work included the usual transfer of supplies, equipment, and trash between the shuttle and space station and various handyman projects. A high priority was repairing the new water purification and recycling system meant to convert urine into drinking water; onboard production would reduce the need to transport and store bags of potable water for ISS crews. After repair, the crew ran a test cycle and took the sample back to Earth for analysis to verify that the recycled water was drinkable.

Something unexpected usually happened during a mission; in this case it was an evasive maneuver. Orbiting space junk is tracked, and the ISS occasionally receives alerts of debris approaching the vicinity. Sometimes a precautionary avoidance maneuver is warranted. Shuttle commander Lee Archambault was called upon to use the docked orbiter's thrusters to give the space station a slight "push" to avoid the potential hazard.

Although they were not all in the same place, thirteen people were briefly in space at the same time during *Discovery*'s return home—seven on the shuttle, three on the ISS, and three in a Soyuz heading there. Only once before, in 1995 during the Shuttle-Mir missions, were so many people in space at once.

When *Discovery* left, the International Space Station was nearly complete. Now the space station architecture was balanced with two pairs of solar arrays at each end of the long girder that, at 335 feet (102 meters) in length, extended longer than a U.S. football field. The last few elements would soon arrive, but with its whole power system the ISS looked finished and could support the full research program for which the station was built.

Seen through the window of the departing orbiter, the International Space Station with its full set of solar arrays seems balanced on the curve of the Earth.

13 : 20 : 54
DAYS   HRS   MINS

221 M
192 NM
(356 KM)

219 | 51.6°

20 : 15
HRS   MINS

3

EDWARDS
AIR FORCE BASE

COMMANDER

- Frederick W. (Rick) Sturckow, USMC, his last of 4 flights, 2 on *Discovery*

PILOT

- Kevin A. Ford, USAF, his only shuttle flight, and an ISS expedition, RMS operator

MISSION SPECIALISTS

- Patrick G. Forrester, USA, engineer, his last of 3 flights, 2 on *Discovery*, RMS operator

- José M. Hernández, electrical engineer, his only flight, RMS operator

- Christer Fuglesang, ESA, physicist, his last of 2 flights, both on *Discovery*, 2 EVAs

- John D. (Danny) Olivas, mechanical engineer, his last of 2 flights, 3 EVAs

ISS CREW (UP)

- Nicole P. Stott, engineer, her 1st of 3 flights, 2 on *Discovery*, 1 EVA

ISS CREW (DOWN)

- Timothy L. Kopra, USA, aerospace engineer, his 2nd flight

Discovery and the International Space Station: A Story in Thirteen Episodes

Leonardo

Although designated an assembly mission, this *Discovery* flight seemed more like a resupply mission because the Leonardo logistics module was the prime payload. *Discovery* had brought this "moving van" full of supplies and equipment to the space station three other times, and then flown a series of missions to deliver and install big exterior structural elements. On STS-128, the focus returned to outfitting the interior, and Leonardo held some large furnishings to be assembled inside the laboratory modules.

Once *Discovery* docked at the end of the Harmony node, crewmembers used the station's robotic arm to lift Leonardo out of the payload bay and berth it to the nadir port on the same node. The logistics module thus temporarily became an ISS appendage. In the course of a week, the combined crews transferred seven tons of equipment and supplies from it into the station, and then loaded Leonardo with research materials and other items for return to Earth. Another couple of tons moved back and forth between *Discovery*'s middeck and payload bay and the space station.

Two refrigerator-size racks of equipment for materials science and fluids research arrived for installation in the U.S. Destiny lab, and a large freezer for experiment samples went into Japan's Kibō lab. An enclosure about the size of a research rack but outfitted as a private sleep

Earth and the International Space Station wrap around the Astronaut Office symbol to represent the continuous human presence in space. The flag of Sweden and Leonardo module in the shuttle's payload bay signify aspects of this mission.

Clockwise from emblem: Kevin A. Ford, José M. Hernández, John D. (Danny) Olivas, Nicole P. Stott, Christer Fuglesang, Patrick G. Forrester, and Frederick W. (Rick) Sturckow. Timothy L. Kopra is not pictured.

compartment was a welcome crew quarters addition. Units like this became necessary as the ISS crew contingent increased from three to six people.

Leonardo also disgorged a fancy new treadmill, nicknamed COLBERT—inspired by cable television personality Stephen Colbert, who had waged a humorous online campaign to persuade NASA to name a module after him. NASA is famous for its acronyms, so the official name became Combined Operational Load Bearing External Resistance Treadmill (COLBERT).

Maintenance tasks around the station dominated three EVAs, chief among them replacing an empty ammonia tank in the truss cooling system and preparing the station for the arrival of a third connecting node. The spacewalkers also removed a materials exposure experiment mounted outside Europe's Columbus lab to return it to Earth for study, and they installed a new experiment to take its place.

The STS-128 mission had several unplanned distinctions. It happened that this was the first shuttle mission to launch at midnight (thus, its launch bridged two days) and the last shuttle mission to land in California. It was the first space mission with two Hispanic American crewmembers and the thirtieth shuttle mission to the International Space Station. It was also the twenty-fifth anniversary of *Discovery*'s first flight. By design, it was the last crew rotation mission; one more ISS crewmember returned on a shuttle, but with the impending retirement of the U.S. vehicles, Russia's Soyuz took over ferry service for ISS crews.

The mission's real significance, though, was equipping the laboratories with advanced research facilities. The space station's incremental buildup had caused some concern about when the orbital research center would be as capable as promised. This *Discovery* mission brought new capabilities and moved the research agenda forward.

Above: Discovery thundered into space as August 28 became August 29, launching at 11:59 p.m. and reaching orbit eight minutes past midnight (EDT).

Left: Nicole P. Stott is surrounded by tons of equipment and supplies packed into the Leonardo logistics module. The combined crews moved all the stowage bags and new racks into the ISS and then repacked Leonardo with items to return.

15 : 2 : 47
DAYS HRS MINS

238 | 51.6°

220 M
191 NM
(354 KM)

20 : 17
HRS MINS
3

KENNEDY SPACE
CENTER

COMMANDER

- Alan G. Poindexter, USN, his last of 2 flights

PILOT

- James P. Dutton Jr., USAF, his only flight

MISSION SPECIALISTS

- Richard A. (Rick) Mastracchio, engineer, his last of 3 shuttle flights, plus an ISS expedition, 3 EVAs

- Dorothy M. Metcalf-Lindenburger, earth sciences educator, her only flight, RMS operator

- Stephanie D. Wilson, aerospace engineer, her last of 3 flights, all on *Discovery*, RMS operator

- Naoko Yamazaki, JAXA, aerospace engineer, her only flight, RMS operator

- Clayton C. Anderson, aerospace engineer, his last of 3 flights, 2 on *Discovery*, 3 EVAs

Superimposed on the gold symbol of the Astronaut Office is the shuttle orbiter bearing the Leonardo Multipurpose Logistics Module on ISS assembly mission 19A. The orbiter appears to be doing the rendezvous pitch maneuver, or "backflip," near the space station.

From left: Richard A. (Rick) Mastracchio, Stephanie D. Wilson, James P. Dutton Jr., Dorothy M. Metcalf-Lindenburger, Alan G. Poindexter, Naoko Yamazaki, and Clayton C. Anderson.

Discovery and the International Space Station: A Story in Thirteen Episodes

Supplies

Discovery flew only once in 2010, the year the Space Shuttle program was supposed to end. Instead, 2011 became the final year, because the space station still needed some work. NASA decided to stock the ISS with as many supplies and spare parts as possible while the big "space trucks" were still in service, for when the shuttles ceased to fly, it would be much harder to send sizable cargo.

STS-131 was the last space station assembly mission flown, and *Discovery* again carried the Leonardo logistics module. Now it was stuffed with more tons of material, including four new experiment racks, another laboratory freezer, an exercise research system, and another crew quarters sleep station. The payload bay held a replacement ammonia tank assembly for the ISS cooling system and other equipment, and another ton or so of bagged items and water occupied the middeck. Most of the equipment was destined for interior assembly of scientific research facilities.

Most of the combined shuttle and space station crew transferred racks and stowage bags from Leonardo and the orbiter into the station and sent returning items in the opposite direction. Meanwhile, the team of spacewalkers and arm operators conducted three EVAs devoted to maintenance tasks. Like the previous *Discovery* crew, this one had to replace a spent ammonia tank located on one of the trusses, a task that consumed almost an entire EVA period. They also replaced a rate gyro unit, installed some crew aids, and finished miscellaneous tasks.

The STS-131 mission marked some firsts and lasts for *Discovery* and the shuttle program. It began as the last night launch of the shuttle era and ended as *Discovery*'s longest mission. This was the last seven-member shuttle crew, the last crew to include rookie astronauts, and the last shuttle crew to include three women—a phenomenon that happened only three times in 135 missions, twice on *Discovery* missions. In fact, four women were in space simultaneously for the first time—*Discovery*'s three plus one on the ISS crew. Also, two Japanese astronauts were in space for the first time, one on the shuttle and one on the space station.

When the seven shuttle astronauts joined the six-person ISS crew for meals and a group portrait, it seemed like a crowded family reunion. After the shuttles were retired, crews would come and go only two or three people at a time in Soyuz capsules, so it would become rare for ten or more to be on the station at once.

Four more times a shuttle would fly to the space station on logistics resupply missions, and a few more features would be added. *Discovery*'s 2011 mission would leave the fully stocked Leonardo module permanently attached to ISS as a warehouse. *Discovery* had first docked with the nascent space station in 2000 to begin equipping it for occupancy; now *Discovery* brought some of the last pieces for its completion.

Richard A. (Rick) Mastracchio (left) and Clayton C. Anderson work to replace an ammonia tank on a segment of the massive ISS truss. Part of the station's Canadarm is in view at the upper left. Part of a module can be seen at the right, and in view behind it is an angled section of radiator panels.

222 M
193 NM
(357 KM)

202 | 51.6°

2

KENNEDY SPACE CENTER

COMMANDER

- Steven W. Lindsey, USAF, his last of 5 flights, 3 on *Discovery*

PILOT

- Eric A. Boe, USAF, his last of 2 flights

MISSION SPECIALISTS

- B. Alvin Drew, USAF, aerospace engineer, his last of 2 flights, 2 EVAs

- Stephen G. Bowen, USN, engineer, his last of 3 flights, 2 EVAs

- Michael R. Barratt, medical doctor, his only shuttle flight, after an ISS expedition, RMS operator

- Nicole P. Stott, engineer, her 3rd shuttle flight, 2 on *Discovery*, and an ISS expedition, RMS operator

Discovery and the International Space Station: A Story in Thirteen Episodes

Finale

Discovery's thirty-ninth and final mission was the 133rd Space Shuttle mission. Only two more would follow, one each by *Endeavour* and *Atlantis*, to bring the shuttle program to an end in 2011. *Discovery* made the most flights to the International Space Station, thirteen since 1999, and this was the thirty-fifth of thirty-seven shuttle-ISS missions. It was also the eightieth docking with the ISS, counting Russian and European spacecraft, too.

Like *Discovery*'s first launch in 1984, this one was plagued by delays. Originally planned to launch in July 2010, STS-133 was postponed repeatedly. *Discovery* moved to the pad in October, but after scrubs and delays for technical and weather reasons it rolled back to the Vehicle Assembly Building in December. *Discovery* finally launched in February 2011 as the first orbiter to make its final flight that year. The extra time permitted some necessary hardware repairs and, because this was a resupply mission, packing as much as possible on board. *Discovery* carried almost twenty tons of hardware and supplies to the space station.

Once again, for the sixth time, *Discovery* delivered the Leonardo module but this time left it there. After being lifted out of the payload bay and handled by both the station and orbiter robotic arms, the newly renamed Permanent Multipurpose Module was attached to the Earth-facing (nadir) side of the Unity node at the center of the space station. Before launch, technicians added protective thermal blankets

Inspired by the late space artist Robert McCall, this design symbolizes *Discovery*'s final launch and return in a long career of missions.

From left: B. Alvin Drew, Nicole P. Stott, Eric A. Boe, Steven W. Lindsey, Michael R. Barratt, and Stephen G. Bowen.

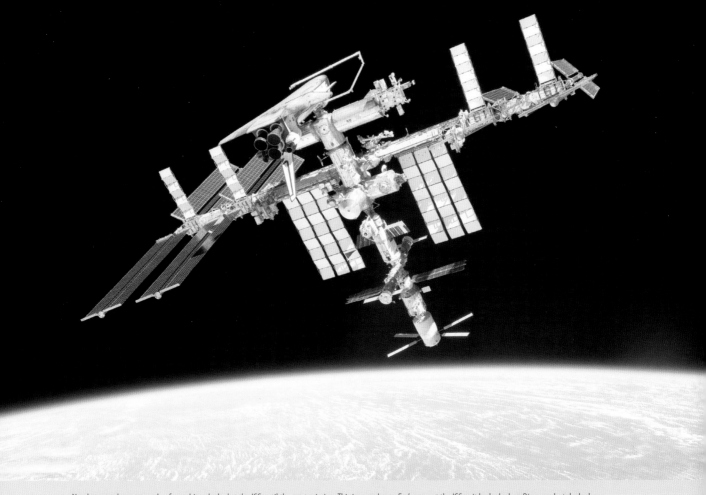

No photographs were made of an orbiter docked at the ISS until the next mission. This image shows *Endeavour* at the ISS as it looked when *Discovery* last docked there. The Leonardo module is at the center of the image, facing in the same direction as the orbiter's engine nozzles. A Russian Progress vehicle is docked at the opposite end of the station.

Discovery approaches the space station for the last time; it brings the Leonardo module and an ExPRESS logistics carrier. The docking adapter, robotic arm, and extension boom are also visible in the payload bay.

STS-133 commander Steven W. Lindsey (in blue) and ISS Expedition 26 commander Scott J. Kelly survey the new ISS storage room, Leonardo.

and micrometeoroid shields as well as fixtures for its permanent attachment as the last major ISS element.

This module arrived full of supplies, equipment, and experiments, and the combined shuttle and ISS Expedition 26 crews promptly spent an entire week unloading and stowing the goods throughout the station. Most of these packages held what ISS crews and research projects would need for the foreseeable future. NASA added two days to the mission to accomplish as much as possible with a double crew, so the module could be restocked with supplies brought on the last two shuttle missions. Thereafter, this module served as a much needed storage room for incoming and outgoing items (and trash) between visits of automated supply craft. Two supply ships from Europe and Japan that were also docked at the ISS were unloaded during this mission.

Discovery also brought a large storage platform to be mounted on the ISS truss. This ExPRESS Logistics Carrier held exterior spare parts—a radiator, a power control unit, and other devices that might be needed as the originals aged. Four such carriers positioned around the station, and other equipment stowed inside the truss bays, held what the station itself might need in the foreseeable future.

The spacewalkers and operators of both arms completed two EVAs to do a number of installation and repair tasks around the station. EVA specialist Steve Bowen, just returned from the STS-132 mission, became the first NASA astronaut to fly on consecutive shuttle missions, and his EVA partner, Al Drew, became the last African American to fly on the shuttle. Coincidentally, all of the crew except commander Steve Lindsey were members of the astronaut class of 2000 (Group 18).

Discovery also delivered a unique payload, the first dexterous humanoid robot in space, Robonaut2. It had a head full of cameras for vision, a torso housing its computers, and two strong arms and grasping hands capable of a wide range

The Leonardo module is seen from an ISS window in its new location as a permanent part of the space station. Also visible is a docked Russian Soyuz spacecraft.

This window view of Stephen G. Bowen on an EVA captures much of the ISS external work environment. The Kibō (left) and Destiny (right) lab modules extend from the obscured Harmony node where *Discovery* is docked, and part of the newly attached Leonardo module is at the right frame. Bowen is working from the end of the station's Canadarm, with the orbiter's combined arm and extension boom behind him.

of motion. In the months ahead, its designers and the ISS crew would test its capabilities as an astronaut assistant in the Destiny lab. Eventually, it might be tried outside to help with EVA tasks.

Before departing, *Discovery* gave the space station a boost, and then the crew prepared for the orbiter's final journey home. Its last twelve years on duty had served to make the International Space Station more commodious for permanent human occupation and more capable for scientific research. When this mission ended, the International Space Station architecture was complete.

But *Discovery*'s job was not yet done. Its final descents made a contribution to aeronautical engineering research. On its last four missions, *Discovery* was modified slightly for a flight dynamics experiment. One tile under the port wing had a wedge-shaped "speed bump" 4 inches long and a quarter to a half inch high—about the size of a pen. Tiles behind it were instrumented to detect when airflow turbulence began and how much heating increased there during reentry. These instances obtained, for the first time, actual data to compare with the computer and wind tunnel modeling that had governed the thermal protection system design for this part of the orbiter.

Upon landing, *Discovery* was retired from the shuttle fleet and moved back into its processing facility to be "deserviced" for future preservation in a museum yet to be chosen. *Discovery*'s retirement signaled the beginning of the end of the shuttle era, a bittersweet occasion for NASA and the shuttle workforce. Longtime shuttle launch director Mike Leinbach put it poignantly: "We wanted to go out on a high note and *Discovery*'s done that."

After posing with "crewmate" Robonaut 2 in the training facility at NASA Johnson Space Center, the STS-133 crew delivered the astronaut-assistant robot to the ISS for testing in the Destiny lab.

Discovery is on final approach to its last landing. With 365 days in space on thirty-nine missions in twenty-seven years, *Discovery* is the champion of the space shuttle fleet.

DISCOVERY THEN AND NOW

4

Discovery at its final landing in 2011 was not exactly the same vehicle as *Discovery* at its first launch in 1984. During its twenty-seven years in service, the orbiter changed inside and out. Safety issues dictated some of the changes, many others led to improved performance, and a few were rather cosmetic. Most of the five hundred or so modifications were not evident except to the engineers and technicians who built and serviced this remarkable machine; they were hidden deep behind the visible surfaces. Yet the Space Shuttle's thirty-year history of 135 flights is a testament to the maintenance and upgrade programs that kept all the orbiters running well and made them better.

At first glance, the most noticeable difference between *Discovery* then and now is its color. No longer pristine white and shiny black, its creamy thermal blankets and streaked, mostly gray tiles were weathered by repeated passage through the furnace of atmospheric reentry. *Discovery* looks like it has been to space and back thirty-nine times, and as a reusable spacecraft, it should naturally show some signs of its age and history.

A closer look at the orbiter's insignia, or livery, reveals intentional exterior changes. *Discovery* entered service sporting on its payload bay and starboard wing the symbol created for the shuttle era—a curvy, stylized NASA logo designed to convey a bold, new look. In the 1990s, NASA retired this symbol, often called "the worm," and returned to its original logo, the round so-called "meatball," to reclaim the heritage of its early years. In 1998, *Discovery* flew for the first time with its new livery. The curvy NASA

Discovery, seen here in the Vehicle Assembly Building, wore the curvy NASA "worm" logo on its first twenty-four missions, 1984–98. *NASA*

121

From STS-95 (1998) on, for its last fifteen missions, *Discovery* bore the round NASA "meatball" logo. *NASA*

logo had disappeared; the round NASA logo appeared on the sides and also replaced the flag and "USA" on the port wing. *Discovery*'s name remained unchanged on the sides of the forward fuselage, but the U.S. flag joined the name on the starboard wing.

Other seemingly superficial features appeared on *Discovery* over the years, most noticeably the addition of two roughly round areas of black tiles on the front of the white Orbital Maneuvering System (OMS) pods. Shuttle workers called them "eyeballs" for obvious reasons, but the black tiles added protection to forward-facing areas that were exposed to more heat than predicted. With time and experience, adjustments occurred in the precise patterns of tiles and blankets on the orbiter, and small labels were added here and there as minor detailing. Along the mid-fuselage sides, the appearance of the payload bay vent doors and hinges subtly changed with decisions about their configuration.

Interior modifications were much more extensive and functional than the relatively few exterior changes. In the 1980s, NASA established a schedule for major maintenance and modification periods, during which each orbiter was taken off flight status for a few months and sent back to "the shop," usually the assembly plant in Palmdale, California. Just as fleet vehicles of any type—aircraft, ships, buses, for example—need occasional overhauls, the reusable orbiters needed more than regular attention between flights. Besides servicing and repair of the normal wear and tear of routine operations, they had to be checked periodically for corrosion, metal fatigue, degraded wiring, and possible systemic problems from repeated use and aging.

Because the orbiters were such complex vehicles and human spaceflight is unforgiving of lapses in safety, NASA cautiously set three years or eight flights as the interval for thorough inspections and heavy maintenance. The interval limit was later extended to 4.5 and then 5.5 years as flights became less frequent. These maintenance periods also made convenient opportunities to upgrade the orbiters, so modifications usually happened at the same time.

In its career, *Discovery* went through three major maintenance and modification periods and a partial one that improved the vehicle and enhanced its performance. These changes tended to be of four types: solutions to problems, upgrades to better equipment, introduction of new technologies, and weight-saving measures. Eventually, all the orbiters except *Challenger* went through the same processes and changes, so *Discovery*'s evolution was not unique.

Discovery's first modification occurred very soon after entering the fleet and flying six missions in quick succession during its first year in service. While the orbiter fleet was grounded after the *Challenger* accident (1986–88), *Discovery* was outfitted with improved nose wheel steering and a crew escape system. The nose wheel steering change addressed unsatisfactory behavior of the original system as the vehicle rolled down the runway. This improvement gave the pilots

Discovery spent nine months of 1995–96 in Palmdale, California, partially disassembled and surrounded by work platforms for its major maintenance and modification. *Rockwell International photo, Courtesy of Dennis R. Jenkins*

better control of the vehicle, especially in crosswinds or with a damaged tire. The crew escape system was a safety measure mandated after the *Challenger* accident. It gave the crew two emergency exit options—an escape pole for parachuting out in flight and an inflatable slide for a rapid exit on the ground.

Discovery's first cyclic major maintenance and modification period occurred in 1992 at Kennedy Space Center. Thorough inspection revealed some corrosion, wiring, and heat shield damage to be repaired, but the main change was installation of a drag chute. Deployed from the base of the vertical stabilizer after touchdown, the parachute helped slow the orbiter and reduce pressure on the brakes as it rolled to a stop. Recurrent severe brake wear and a blown tire had plagued earlier missions and forced a policy of landing in California until these problems were solved. Redesigned brakes and use of a drag chute addressed that issue. In an ongoing effort to reduce vehicle weight

A U.S. Air Force parachutist tests the new crew escape pole installed in a C-141 aircraft. It would allow a shuttle crew to bail out in an emergency under suitable flight conditions. *NASA*

to permit carrying more payload mass, *Discovery* also received some lighter weight crew equipment—seats, galley, racks, and lockers—in the middeck.

Discovery's second cyclic overhaul and modification occurred in Palmdale in 1995–96. The major changes this time added a new capability—orbital docking with a space station, first the Russian Mir and later the International Space Station. Workers removed the internal airlock, installed the external airlock behind the hatch to the payload bay, and mounted the orbiter docking system on top of the airlock. Other upgrades included better payload bay lighting and tire pressure monitors. In another weight-saving measure, thermal protection pads replaced many quilted thermal blankets on the forward fuselage, upper wings, and payload bay doors. Engineers looked everywhere to shave weight. In an effort to save ounces, if not pounds, reflective aluminum foil tape in the wheel wells was replaced with aluminized Kapton tape.

Discovery's last major maintenance and modification period occurred during the flight hiatus after the early 2003 loss of *Columbia*. *Discovery* had just rotated

out of service in late 2002 to receive the Multifunction Electronic Display System (MEDS), or "glass cockpit," upgrade and the anti-micrometeoroid orbital debris radiator upgrade, plus a number of other less noticeable performance-enhancing modifications.

The glass cockpit was undoubtedly the most glamorous and impressive new technology added to the orbiters. It transformed the "dashboard" of the cockpit from an array of quaint mechanical meters and gauges and three small monochrome screens to a bank of nine large, flat-panel, multicolor, highly interactive liquid crystal display monitors and two more in the aft flight deck. The softly glowing screens and backlit switches in the darkened cockpit looked like those in a fantasy starship, but they already were standard features in the most modern military and commercial aircraft. The shuttle was now ready for the twenty-first century.

The glass cockpit was not for show, however. The genuine purpose of presenting and integrating key flight data electronically in easy-to-read graphics was to aid pilots' understanding and decision making, enabling them to work smarter and more efficiently. This technology also weighed less and used less power—added advantages in a spacecraft environment where every pound and watt were husbanded.

Discovery also received a somewhat exotic although not readily visible upgrade to better protect its radiator panels from micrometeoroid strikes. The radiators lining the payload bay doors were exposed during the shuttle's time in orbit to rid the vehicle of excess heat from the onboard equipment. Millions of bits of rocket and satellite debris swirl around Earth at velocities comparable to the shuttle's, and collision with even a tiny fleck of paint or metal can have the effect of a bullet. Because post-flight inspections revealed "bullet holes" and craters from debris impacts, NASA decided to add protective strips of aluminum over the most vulnerable parts of the radiators to thicken the surface just enough to reduce the risk of damage.

Opposite: Dressed for clean room work on flight hardware, technicians install the external airlock in *Discovery*'s payload bay. The bay liners have been removed for access to the underlying mechanical, fluid, and electrical systems. *NASA*

Above: *Discovery* rolls to the end of a 1997 mission, shortly after being equipped with the drag chute upgrade. *NASA*

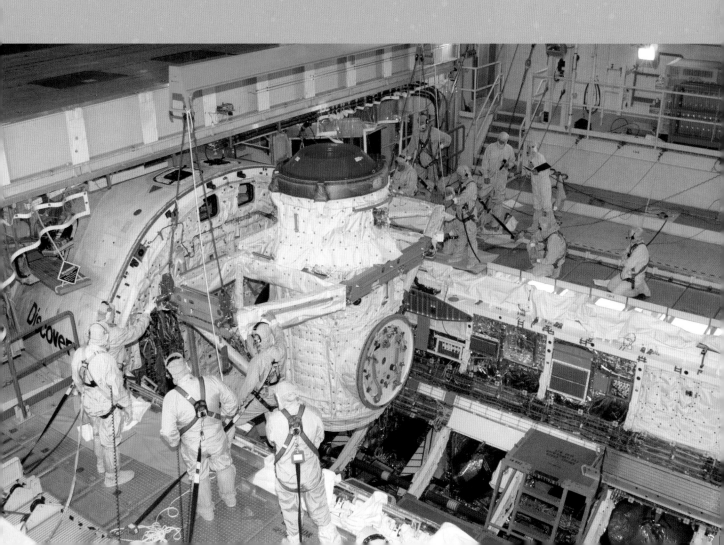

The glass cockpit replaced about forty old meters and gauges with a state-of-the-art computer graphics video system. *Discovery* received the cockpit upgrade just before the 2005 return-to-flight mission. *NASA*

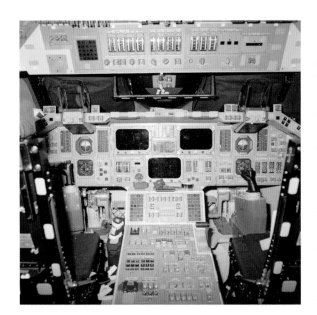

The original cockpit was organized around electromechanical flight instruments linked to the orbiter's general purpose computers and three cathode ray tube (CRT) data display units—a quite advanced system in the early shuttle era. *NASA*

By coincidence rather than plan, *Discovery* became the return to flight orbiter after both the *Challenger* and *Columbia* tragedies. *Discovery* happened to be the vehicle most recently maintained and upgraded—in 1988 with regular and post-accident improvements and in 2005 with the glass cockpit and a number of modifications for increased safety. Among the post-*Columbia* changes, *Discovery* first received the Orbiter Boom Sensor System (OBSS), a straight extension to the orbiter's robotic arm that was equipped with a laser, cameras, and sensors to inspect the entire vehicle for damage that might hinder its safe return. *Discovery* also was equipped first with wing leading edge detection sensors to record any impact there during liftoff and ascent. Called up because it was most ready to fly, *Discovery* twice put American astronauts, and through them the United States, back in space.

Discovery and the other orbiters received approximately five hundred upgrades during the Space Shuttle program. Hardly any system was not touched.

New computers, brakes, tires and wheels, avionics and navigation devices, auxiliary power units, fuel cells, thermal protection, other equipment, and some structural changes kept the orbiters updated and at peak performance. Change was less often a response to flaws than to advantages from improved technologies. Change kept the reusable shuttle always capable for its missions as America's premier spacecraft.

Making the orbiters safer and more reliable—that was the rationale for modifications and upgrades. Also, these changes usually made the vehicles easier to maintain. If hardware proved unreliable, a better replacement reduced the maintenance workload. As equipment aged or failed, and it became harder to find spares or make repairs, a modification could solve that problem. The shuttle orbiters were a continually improving work in progress, tended to by a workforce committed to keeping the fleet ready and safe for flight.

After the loss of *Columbia* and President George W. Bush's announcement that the Space Shuttle program would end upon full assembly of the International Space Station, a number of editorial cartoonists poked fun at this remarkable vehicle. They drew the orbiter as a patient in a hospital bed or a wheelchair or hobbling along behind a walker or using a cane, surrounded with pill bottles and intravenous bags and other symbols of infirmity. These portrayals were a sad disservice to the orbiters and their keepers as well as a misperception of their remaining vigor.

Although *Discovery* was older and flew more often than the other orbiters, it was hardly geriatric. By some measures, it was safer and more ready to fly for its last mission than for its first. When forced into early retirement, it was time for its next major maintenance and modification period. Had that happened, it might have emerged with a "smart cockpit" to take over some of the pilots' more routine tasks, an orbiter health system to self-diagnose some of its own problems, or a variety of new technologies for continued space transportation. Instead, *Discovery*'s final servicing prepared it to enter the Smithsonian National Air and Space Museum for permanent display as the champion of the Space Shuttle fleet.

The jointed Canadarm remote manipulator arm, with the straight orbiter boom sensor system attached, rests over the payload bay. The combination is long enough and flexible enough to reach all parts of the orbiter for inspection. When not in use, the extension is stowed along the opposite (starboard) edge of the payload bay. *NASA*

5

DISCOVERY's FINAL MISSION

On the thirtieth anniversary of the first Space Shuttle launch, NASA Administrator Charlie Bolden stood onstage before the orbiter processing facility in Florida to make an announcement. But first he gave a farewell address to the shuttle and praised its workforce. *Discovery* had just returned from its final mission, and the shuttle era was now coming to an end. A veteran pilot and commander of missions on three of the orbiters, this retired U.S. Marine Corps general clearly felt affectionate pride for these remarkable ships.

Then he broke the news that his crowd and audiences at museums across the country nervously anticipated: the fate of the orbiters. For two years, NASA had vetted sites that applied to acquire an orbiter for permanent display. The process grew quite competitive, as schoolchildren and other citizens flooded the agency with letters and petitions and political figures at all levels urged the merits of their favored site. There had been rumors but no leaks of NASA's decision, and hardly anyone knew what Bolden might say.

First, *Atlantis* would stay at Kennedy Space Center, home port of the Space Shuttle fleet.

Next, *Endeavour* would go to the California Science Center in Los Angeles, near the design and assembly plants where the orbiters originated and the desert landing site where they came and went.

Discovery, the most-traveled orbiter, would go to the Smithsonian National Air and Space Museum to join the most historic aircraft, spacecraft, and national treasures.

Discovery descends past the National Air and Space Museum's Steven F. Udvar-Hazy Center on final approach to touchdown at Dulles International Airport. *Photo by Dane Penland, National Air and Space Museum (NASM 9A09800)*

NASA marked the thirtieth anniversary and end of the Space Shuttle program with an emblem that commemorated the five orbiters and fourteen lost crewmembers as stars in a design suggesting upward thrust from Earth orbit into outer space. *NASA*

And finally, the test orbiter *Enterprise* would move from the Smithsonian to the Intrepid Sea, Air, and Space Museum in New York City.

Bolden's announcement set in motion the last year of the Space Shuttle program, which was spent delivering the orbiters to their new homes and phasing out the equipment, facilities, and most of the personnel that had made the past three decades of U.S. human spaceflight so successful. President George W. Bush had announced in 2004 that the shuttle program would come to an end; this April day in 2011 marked the ceremonial beginning of the end.

NASA Administrator Charles F. Bolden Jr., announces the orbiters' final destinations at a Kennedy Space Center ceremony on April 12, 2011, the thirtieth anniversary of the first Space Shuttle launch. *Discovery*'s aft end towers behind him in an orbiter processing bay. *NASA*

Discovery's OMS pods and forward reaction control system module were sent to NASA's test facility in White Sands, New Mexico, to be detoxified. Here, the port OMS pod returns clean for reinstallation. *NASA*

Retiring the orbiters was not a simple matter. Each one underwent about a year of "deservicing" and "safing" before delivery to its permanent display site. No matter how good the vehicle looks on the outside, an orbiter's inner workings are polluted with noxious, even toxic, chemicals. The orbiter also carries explosive devices and some components held under high tension. NASA was obliged to identify and neutralize such hazards to the greatest extent possible, short of gutting the vehicles, so the orbiters could be safely left on public display forever. For the orbiter processing teams, this requirement was an unprecedented challenge, in part a reversal of their normal workflow.

Discovery went through the process first, and removal of toxic substances—the highest priority—required major surgery. The propulsion and environmental control systems circulated monomethyl hydrazine, nitrogen tetroxide, hydrazine, ammonia, Freon, and other fluids from tanks through scores of valves and miles of lines throughout the vehicle. Although these systems were drained and purged between missions, any toxic residue embedded in the equipment could not be eliminated completely.

Because slow outgassing over time might accumulate and leak out, NASA decided to remove certain elements to avoid the risk of possible public exposure.

The Orbital Maneuvering System (OMS) pods and the module housing the forward reaction control system thrusters—the most contaminated parts of the orbiter—were gutted. Although *Discovery* and the other retired orbiters appear whole, those areas are almost empty shells. Other noxious chemicals or materials, such as asbestos and beryllium that were enclosed and inaccessible, were carefully documented.

The next priorities were to remove or release stored energy. Explosive pyrotechnic devices are tucked into places around the orbiter to force a mechanical operation, such as lowering the landing gear, if it does not occur on its own. Those were removed. Other components with stored mechanical energy, such as the crew escape system and other spring-loaded or rapidly inflatable equipment, were released and safed.

Some parts of the orbiters were surgically removed for reasons other than safety. NASA decided to retain the Space Shuttle's main engines for possible use in future launch vehicle development. The shuttle's engines remain the most sophisticated and most efficient, although not the most powerful, large liquid propellant engines on the market and still have a useful shelf life. All main engines and their nozzles were removed and returned to inventory. Technicians refurbished test nozzles and attached them to a block

Without engine nozzles, OMS pods, and part of its nose, *Discovery* moves from temporary storage in the Vehicle Assembly Building back to the orbiter processing facility for further work. *NASA*

of metal placed in the aft compartment amid the propellant feedlines and pressurization tanks. Again, the orbiters look complete from the outside but are missing three major "organs" in the aft.

NASA programs and labs eyed a variety of other orbiter components for various purposes, and some requests for shuttle hardware removal were approved. However, NASA and the National Air and Space Museum agreed to exempt *Discovery* from most of those discretionary extractions. Both organizations recognized the importance of preserving one orbiter as intact as possible to serve as the Space Shuttle historical reference vehicle. As a result, *Discovery* underwent fewer modifications upon retirement and remained more complete than the others—not exactly in as-last-flown condition, but close.

Meanwhile, as *Discovery* moved through deservicing, the museum and NASA began to prepare for another complex operation: delivering *Discovery*, exchanging it for *Enterprise*, and transferring *Enterprise* to its new home. Orbiters were ferried on top of the shuttle carrier aircraft, a modified Boeing 747 jetliner, and they were lifted up and down in a permanent crane fixture in Florida or California at the termini of ferry flights.

The logistics for delivering a retired orbiter were daunting, because the technical team had to do the lifts "in the field" without their specialty crane setup. That had happened only rarely in the 1970s and early 1980s, when *Enterprise* was relocated to places other than the normal termini and when *Columbia* (STS-3) had to land in White Sands, New Mexico. Procedures existed, but had never been used, for sending an orbiter back from an emergency landing site overseas. No current workers had experience rigging and loading or unloading an orbiter in the field, so procedures had to be reconstructed, adapted, and rehearsed.

Fortunately, the first orbiter delivery—*Discovery*—was to an airport located next door to the orbiter's destination, the National Air and Space Museum's Steven F. Udvar-Hazy Center in suburban Virginia. The Metropolitan Washington Airports Authority and Dulles International Airport managers cooperated with NASA and United Space Alliance (USA), its shuttle servicing contractor, to find an on-site location where the carrier aircraft could park for the orbiter demating and mating activities. A very large concrete area used for deicing aircraft was available after the snow season ended. They permitted NASA to set up and occupy for a month a complex of huge cranes, support towers, supply and office trailers, and support equipment for a technical team that ranged from forty to eighty people.

Looking ahead to transporting *Enterprise* to New York, NASA decided to prepare it for flight well in advance to have that work out of the way before *Discovery*'s arrival. *Enterprise* had last "flown" in 1985 when it was delivered to the museum. It had spent more than twenty-five years in storage or on static display. The vehicle needed a thorough inspection to ensure that it was still ferry-worthy; if corrosion or any other damage had weakened its structure and moving parts, moving *Enterprise* would be problematic. Also, *Enterprise* was not exactly the same as the newer orbiters, and some of its features did not match up properly with the tools and attachment fixtures now used for maintenance and ferry flight. NASA-USA teams

modified *Enterprise* where necessary, put it in flight configuration, weighed it, and determined its center of gravity. They also installed the original OMS pods and tailcone from the vehicle's test flights to enhance its historic authenticity. *Enterprise* was ready to go.

The museum also faced a logistics challenge. *Enterprise* stood exactly where *Discovery* would go, in the center of the James S. McDonnell Space Hangar. Scores of other artifacts surrounded it on the floor, in display cases, and suspended overhead. To make the orbiter exchange, museum staff had to remove and put in temporary storage everything aft of the orbiter's wingspan. That gave the NASA-USA team adequate room to work on *Enterprise* and cleared the way for *Enterprise* to be rolled back through the hangar door onto the museum's haul way, and for *Discovery* to enter. Planning began immediately to fit these relocations into the museum's busy schedule.

Orbiters were hoisted on and off the shuttle carrier aircraft in a mate-demate facility in Florida or California, the end points of their ferry flights. *Discovery* was placed on the Boeing 747 in this device for its last ferry flight, but it was off-loaded with mobile cranes in wide-open space at Dulles International Airport. *NASA*

The demate worksite at the airport occupied about 200,000 square feet (18,580 square meters). A convoy of some one hundred trucks brought the cranes, masts, lift fixture, tools and equipment, and trailers needed for the job. The huge white crane came in pieces on fifty trucks and was assembled in place. *Courtesy of Dennis R. Jenkins*

ORBITERS ON DISPLAY

The team who prepared and transferred the orbiters to their final destinations adopted this logo showing the carrier aircraft, two cranes, and lifting fixture needed for delivery and demate operations. *Original art by Tony R. Landis*

The museum's other challenge was a pleasure: planning a welcome celebration worthy of the champion of the Space Shuttle fleet and the nation's capital. That effort began shortly after the announcement that *Discovery* would go to the Smithsonian. At the top of the list along with a safe delivery, NASA and the museum hoped that the fly-in would include a fly-over of Washington, D.C., the most restricted public airspace in the country. Requesting and coordinating the fly-over involved representatives of the museum, NASA, Dulles International Airport, the Metropolitan Washington Airports Authority, Federal Aviation Administration, Department of Homeland Security, and local law enforcement bureaus.

They gained provisional approval for a fly-over, dependent on weather, timing, and security considerations. Dulles International Airport identified a low air traffic time slot that would be suitable for the fly-in and landing. Because final approval came only after *Discovery* was airborne, NASA and the museum stoked anticipation by issuing general public advisories to be prepared to "spot the shuttle" until they could make an announcement of the arrival time.

On Tuesday, April 17, 2012, the weather cooperated and word went out through all media. From 10 o'clock until almost 11 in the morning, residents and tourists stopped what they were doing and stepped outside to watch the piggybacked air and space craft. The 747 flight crew made a first pass over the Udvar-Hazy Center to the delight of the crowd that filled the parking lot and observation tower. As the airports held other air traffic, the NASA plane turned east and flew "low and slow" along the Potomac River, past Reagan National Airport, over the museums on the National Mall and NASA Headquarters nearby, around the Capitol, west toward the White House and Washington Monument, and back to Dulles. Facebook, Twitter, and traditional media lit up with images and sighting reports.

Observers at the Udvar-Hazy Center assumed that *Discovery* was ready to land when it reappeared, but the 747 landing gear were not down. Off the plane went on another loop around the District and the Mall,

this time ranging farther to pass over NASA's Goddard Space Flight Center in Maryland and a Federal Aviation Administration facility in Virginia. Arriving at Dulles the third time, the shuttle carrier aircraft touched down and then brought *Discovery* on a brief parade past the terminals before moving to the worksite.

After an hour or so of welcome remarks and media attention, dignitaries and onlookers cleared the area so the technical team could get to work demating *Discovery* from the 747. They had a tight schedule to get the orbiter back on the ground and rolled over to the museum for a formal welcome ceremony on Thursday, April 19. The plan called for preliminary staging work in the afternoon, a good night's rest, and a full day to demate the orbiter on Wednesday.

Everything thus far occurred right on schedule, and Wednesday morning started well, although it had

Discovery left Florida for the last time at daybreak. Three hours later, the shuttle carrier aircraft arrived in the Washington, D.C., area. *NASA*

rained overnight. However, in the afternoon the wind picked up just a few knots too much to do the delicate lift. The orbiter would be suspended between two cranes and steadied by guy wires connected to four tall masts. This position was too precarious to hold in wind, so the lift had to be postponed until the wind decreased, customarily in the evening.

The operations chief called a halt and sent everyone to their hotels to rest, with a tentative plan to resume work at 10 p.m. By then the wind abated, but a light rain began. Nevertheless, the crew worked through the night, doing a complicated job for the first time in the dark and rain, without a problem. By 7 o'clock the next morning, *Discovery* was waiting at the gate in the fence that marked the border of the airport and museum properties. Just as the shuttle workforce

A study in contrasts: Twentieth-century air and spacecraft pass within view of a nineteenth-century Washington landmark, the Smithsonian Institution Castle. *James DiLoreto for NASA/Smithsonian Institution*

Discovery makes the first of two passes over Washington, D.C. *Robert Markowitz for NASA/Smithsonian Institution*

Within view of the Dulles International Airport control towers and terminal buildings, *Discovery* waits for wind and rain to subside so the delivery team can off-load the orbiter. *Photo by Dane Penland, National Air and Space Museum (NASM 2012-01849)*

did for thirty years of spaceflight operations, this crew gave their best effort to this mission.

Thursday, April 19, dawned sunny, and the Udvar-Hazy Center began to fill early with people who wanted to witness *Discovery*'s formal arrival at the museum and sign the "Welcome Discovery" banner. One area of the parking lot filled with media trucks from local, national, international, and cable networks, and journalists roamed the area interviewing and photographing.

The museum's education staff organized a "Welcome Discovery" fair of exhibits, demonstrations, and activities where visitors for the next few days learned about rockets, spacesuits, space food, and other topics related to the shuttle and spaceflight. The museum shop featured shuttle items, the café promoted shuttle themed items, and the theater showed *The Dream Is Alive*, the first IMAX® film shot in space, some of it on a *Discovery* mission. The lawn behind the building, along the haul way from the airport, filled with people in a festive mood.

The official outdoor ceremony began with a processional by the U.S. Marine Corps Drum and Bugle Corps in recognition of the centennial of Marine Corps aviation. *Discovery* rolled into view, escorted by thirty *Discovery* commanders, mission specialist astronauts, and the NASA-USA work crew who delivered the orbiter, and came to a stop behind the stage, nose to nose with *Enterprise*, which was already waiting there. After Washington's resident opera star, Denyce Graves, sang the national anthem and a presentation by the Marine Corps Color Guard, the director of the National Air and Space Museum, Gen. John R. Dailey, USMC (ret.), presided over a brisk program.

Working in rain and darkness, the NASA-USA team demate *Discovery* "in the field," an event that had not happened since 1985 when *Enterprise* was delivered to the museum. *Courtesy of Dennis R. Jenkins*

A short video recapped *Discovery*'s storied career before NASA Administrator Maj. Gen. Charles F. Bolden Jr., USMC (ret.), verbally transferred the orbiter to the museum. Two Smithsonian officials accepted, Secretary Wayne G. Clough and Board of Regents Chair Dr. France A. Córdova. Former Mercury astronaut and Senator John Glenn, who returned to space on *Discovery* in 1998 and like Generals Dailey and Bolden is a former USMC aviator, spoke about the history of spaceflight. Then the moment came to culminate the transfer by signing ceremonial documents. The festivities ended with video highlights of the arrival fly-over.

Late in the afternoon, the physical orbiter exchange and transfer into the space hangar began. *Enterprise* had been rolled outside in the morning and brought into position behind the stage. Now it was pushed back past the space hangar to allow *Discovery* to roll forward and turn into the hangar. In less than an hour, *Discovery* was positioned inside on public display. Then *Enterprise* was rolled forward again to the gate in the boundary fence and on to the airport worksite to be loaded atop the carrier aircraft for its final journey.

The delivery team stayed for another week to upload *Enterprise* and dismantle their encampment at the airport. A smaller crew spent about two weeks in the space hangar with *Discovery* up on jacks so they could remove the tail cone used for ferry flight, secure the landing gear, install the OMS nozzles, adjust the main engine nozzles out of ferry flight position, and do some interior detailing. Two experts flew in to assemble the Canadarm, which the museum decided to display outside *Discovery* to be seen and appreciated. After this work was complete, museum staff installed the guard rail and exhibit panels around *Discovery* and began the long process of returning all the other artifacts and displays.

Since taking up residence at the Udvar-Hazy Center, *Discovery* became a star attraction. More than 1.5 million people came during its first year at the museum. Visitors often approach the information desk near the

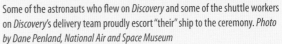

Some of the astronauts who flew on *Discovery* and some of the shuttle workers on *Discovery*'s delivery team proudly escort "their" ship to the ceremony. *Photo by Dane Penland, National Air and Space Museum*

welcome
Discovery

Smithsonian
National Air and Space Museum
Steven F. Udvar-Hazy Center

Launching at the Steven F. Udvar-Hazy Center this Spring
for details, visit www.airandspace.si.edu/discovery

Above: The stage is set for a historic meeting of the shuttle test vehicle, *Enterprise*, and the longest serving, most travelled orbiter, *Discovery*. *Carolyn Russo for NASA/Smithsonian Institution*

Left: The museum welcomed the community to celebrate the springtime arrival of its newest acquisition. *National Air and Space Museum*

Right: Senator John Glenn gives the keynote address, flanked by other program participants and many of *Discovery*'s commanders. *Photo by Dorothy Cochrane, National Air and Space Museum (NASM 2012-01566)*

Below: The orbiter transfer is completed as *Discovery* enters the space hangar and *Enterprise* moves past, heading to the airport for its last flight. *Photo by Dorothy Cochrane, National Air and Space Museum (NASM 2012-01556)*

entrance saying, "Where is the shuttle? We want to see *Discovery*." It is actually visible from the entrance, at least the tall vertical stabilizer and the rounded OMS pods, and the payload bay and nose come into view as one approaches. For many, standing face to face with *Discovery* is a surprise; it is so much larger than a Space Shuttle looks on television or in photos. People marvel at its features, especially its weathered tiles and blankets as well as its sheer size compared to the Mercury, Gemini, and Apollo capsules across the aisle.

In the Museum, *Discovery* often becomes the backdrop for talks about spaceflight by docents, staff, or visiting astronauts. It is the focus for a variety of educational materials and programs developed by the on-site educators. NASA has brought Congressional staff to see *Discovery* on technical familiarization briefings. It is a popular setting for news segments and live broadcasts and for documentary films. *Discovery*'s final and perpetual mission is now education and inspiration. It stands as an icon of the achievements of the Space Shuttle era and the concept of routine spaceflight.

Now and always, *Discovery* remains the champion of the Space Shuttle fleet, and at the Smithsonian it will be preserved forever as a venerable American treasure.

With *Discovery* on jacks for final configuration work, the NASA-USA team gathers for a group portrait. Some were laid off within days, as soon as their last job in the shuttle program phase-down was done. *Photo by Dane Penland, National Air and Space Museum*

Discovery is central to the museum's display of twentieth-century space history in the James S. McDonnell Space Hangar. *Photo by Dane Penland, National Air and Space Museum (NASM 2013-02577)*

GLOSSARY

ACES	Advanced Crew Escape System pressure suit
ACTS	Advanced Communications Technology Satellite
ASTRO-SPAS	Astronomy Shuttle Pallet Satellite
ATLAS	Atmospheric Laboratory for Applications in Space
ATV	Automated Transfer Vehicle
Canadarm	Remote Manipulator System robotic arm, made in Canada
CSA	Canadian Space Agency
CRISTA	Cryogenic Infrared Spectrometers and Telescopes for the Atmosphere
DOD	Department of Defense
ESA	European Space Agency
EVA	Extravehicular Activity, spacewalk outside a spacecraft
GAS	Get Away Special canister for small experiments in the payload bay
GPS	Global Positioning System
HERCULES	Hand-Held, Earth-Oriented, Real-Time, Cooperative, User-Friendly, Location-Targeting and Environmental System; a video camera with GPS-like navigation capability
HOST	Hubble (Space Telescope) Orbiting Systems Test
ISS	International Space Station
IUS	Inertial Upper Stage, solid propellant booster rocket
JAXA	Japan Aerospace Exploration Agency
JEM	Japanese Experiment Module, Kibō
LES	Launch-Entry Suit
LIDAR	Light Detection and Ranging
LITE	LIDAR In-Space Technology Experiment
Miles	One statute mile (5,280 feet) equals 0.87 nautical miles or 1.61 kilometers. One nautical mile (6,076 feet) equals 1.15 statute miles or 1.85 kilometers.
Mir	Russian space station, 1986–2001

MMU	Manned Maneuvering Unit propulsion backpack
MPLM	Multipurpose Logistics Module (Leonardo, Raffaello)
NASA	National Aeronautics and Space Administration
NASDA	National Space Development Agency of Japan; in 2003, became JAXA
OBSS	Orbiter Boom Sensor System
OMS	Orbital Maneuvering System, OMS pod
ORFEUS	Orbiting Retrievable Far and Extreme Ultraviolet Spectrometer
PAM	Payload Assist Module, solid propellant upper stage booster rocket
RMS	Remote Manipulator System robotic arm, Canadarm
ROMPS	Robot Operated Materials Processing System
RSA	Russian Space Agency, also called Russian Federal Space Agency and Roscosmos
SAFER	Simplified Aid for EVA Rescue
SDIO	Strategic Defense Initiative Organization
SPARTAN	Shuttle Pointed Autonomous Research Tool for Astronomy
SPAS	Shuttle Pallet Satellite
SPIFEX	Shuttle Plume Impingement Flight Experiment
STS	Space Transportation System
TDRS	Tracking and Data Relay Satellite
TOS	Transfer Orbit Stage, solid propellant upper stage booster rocket
UARS	Upper Atmosphere Research Satellite
USA	United Space Alliance, in Chapter 5
USA	United States Army
USAF	United States Air Force
USCG	United States Coast Guard
USMC	United States Marine Corps
USN	United States Navy
WINDEX	Window Experiment